Icons of Invention

American Patent Models

Edited by Barbara Suit Janssen

A traveling exhibition produced by the
National Museum of American History,
Smithsonian Institution
Washington, D.C.
1990

Cover photograph by Richard Strauss

Sewing Machine
Elias Howe, Jr.
September 10, 1846
Patent No. 4750

Library of Congress Cataloging-in-Publication Data

Icons of invention: American patent models / edited by
Barbara Suit Janssen.
p. cm.
"A traveling exhibition produced by the National Museum of American History, Smithsonian Institution, Washington, D.C."
ISBN 0-929847-04-0
1. Models (Patents)—United States—Exhibitions. I. Janssen, Barbara Suit, II. National Museum of American History (U.S.)
T223.U3I27 1990
609.73—do20 90-9743
CIP

Introduction

In his famous work *Democracy in America*, French traveler and political observer Alexis de Tocqueville described Americans of the 1830s as innovative, energetic, speculative, and optimistic. De Tocqueville characterized America as a "land of wonders," rich in natural resources. True, its transportation system, manufacturing capacity, and financial machinery remained rudimentary. But these conditions were changing rapidly. He saw a people that were "enterprising, fond of adventure, and, above all, of novelty." America struck him as a land in which everything would be kept "in constant motion."

By reforming the national patent system in 1836, the American government both reflected and encouraged the enterprising, innovative spirit witnessed by de Tocqueville. The Patent Office was elevated to the status of a bureau within the State Department. Expert examiners were appointed to evaluate each application's "novelty, originality, and utility." The law required each applicant to "furnish a model of his invention . . . of a convenient size to exhibit advantageously its several parts." Finally, a monumental Patent Office building was to be constructed, where these patent models would be displayed, the aim being to stimulate inventive creativity and advertise American ingenuity. Government officials in Washington felt certain that an exhibition of patent models, along with samples of resources, examples of manufactures, and historic artifacts, would help instill a patriotic sense of America's destiny.

The first of the Patent Office's four wings was completed in 1840, and in four decades the building filled up with models as quickly as new space was provided. By the time the requirement for submitting patent models was dropped in 1880, nearly a quarter-million of them had been deposited. Later, because space was needed for other purposes, many of these models were sold to private collectors. But the Smithsonian Institution played a key role in making certain that a significant proportion of them would remain in the public domain, as immediate and tangible evidence of American ingenuity in the nineteenth century.

Founded in 1846 as a result of a $500,000 bequest to the United States from English scientist James Smithson, the Smithsonian Institution had emerged as an important educational and research institution. In the 1850s it began to assume the role of a national museum as well. It started adding patent models to its collections in the 1880s and has kept on acquiring them ever since.

The National Museum of American History—one of several museums that are now part of the Smithsonian Institution—currently holds nearly 10,000 patent models, the largest and most comprehensive collection in any public institution. We feel that this collection conveys a sense of the inventive spirit that motivated nineteenth-century Americans, people who were seeking to realize their full potential in a bountiful and challenging land.

Douglas E. Evelyn
Deputy Director
National Museum of American History

ACKNOWLEDGMENTS

I wish to thank the many people who accomplished the projects described here for the Smithsonian. On the Museum's staff, Barbara Janssen managed and curated the exhibitions and catalogs, supported by curators and specialists who contributed essays to this catalog, provided information on the models, and assisted in organizing the exhibitions. Essays for the catalog were written by Kendall Dood, Robert Post, John White, Roger White, Rodris Roth, Anne Serio, Bernard Finn, William Worthington, Elizabeth Harris, Kate Carlisle, Carlene Stephens, Pete Daniel, Rita Adrosko, Barbara Janssen, Ellen Hughes, and Gary Sturm. Robert Post, Robert Selim, and Nancy Brooks edited this catalog.

Additional support was provided by Anastasia Atsiknoudas, Betty Sharpe, Nance Briscoe, Tom Crouch, Francis Gadson, Anne Golovin, Pat Harman, Michael Harris, Peggy Kidwell, Peter Liebhold, Arthur Molella, Stanley Nelson, Jennifer Oka, Eugene Ostroff, Terry Sharrer, David Shayt, Elliot Sivowitch, John Stine, Susan Tolbert, Helena Wright, and Kay Youngflesh. Special thanks also go to Martin Burke, Nikki Horton, Scott Odell, Beth Richwine, Suzanne Thomassen-Krauss, and W. David Todd of the Division of Conservation, for their advice and conservation of the models. Complexities of packing and shipping a traveling exhibition were ably met by Martha Morris, our registrar, and her chief assistant, Catherine Perge. Additional assistance was provided by Edward Ryan, Cheryl Washer, Rosemary Wilcox, Melodie Kosmacki, Nancy Card, Diane Gatchell and Rosemary DeRosa. Administrative and secretarial support was provided by Debora Scriber, Sarah Rollins, and Joan Lashley. The Japanese exhibition benefited from the expertise and assistance of Joseph Shealy, Contracts Specialist, and Elizabeth Driscoll, SITES Exhibition Coordinator.

The catalog photographs from the Smithsonian Office of Printing and Photographic Services were primarily the work of Richard Strauss. Support was also provided by Laurie Minor, Richard Hofmeister, Mary Ellen McCaffrey, and John Dilibar.

We thank the Association for Japan-U.S. Community Exchange (ACE) for making the Japanese exhibition and catalog possible—for providing the resources and working with us to bring it about. In association with ACE, the following sites were selected for the exhibition tour from July 1, 1989, to December 3, 1989:

- Communication Museum, Tokyo
- Sogo Department Store, Yokohama
- Sogo Department Store, Matsuyama
- Sogo Department Store, Hiroshima
- Sogo Department Store, Kobe
- Saito Hoonkai Museum of Natural History, Sendai
- Nagoya Municipal Science Museum, Nagoya

Gracious hospitality and organization of the many parts of the exhibition were administered by Kyoko Ito and Kimiko Hirotsu. Kouichi Sasaki, a noted international exhibition designer, dealt with the complexities of a traveling exhibition. His understanding of American innovations and how they could be displayed is greatly appreciated. The Japanese catalog was designed by Tsuyoshi Fuchigami, edited by Yutaka Imamura of the Bijutsu Shuppan Co. Ltd., and printed by the Dainihon Printing Co. Ltd. Many of the patent models were photographed by Norihiro Ueno. We are especially grateful to the following Japanese sponsors: Nippon Telegraph and Telephone, Kokusai Denshin Denwa, Pioneer, Nomura Securities, and the Electric and Power Companies of Tokyo, Kansai, Chubu, Tohoku, Chugoku, and Shikoku. Without their support, this exhibition and catalog would not have occurred. Jeffrey Stann in the Smithsonian Office of Membership and Development and Hanako Matano in Tokyo assisted the Smithsonian in its arrangements with ACE.

To commemorate the Patent Act of 1790, we have collaborated with the National Portrait Gallery to produce the exhibition, "Icons of Invention." Special thanks go to Allen Fern, Director; Beverly Cox, exhibition curator; and Nello Marconi, exhibition designer, for orchestrating the exhibition. Support for this exhibition and catalog were provided by the Foundation For A Creative America. They were ably represented by Ellen Cardwell, Covener for the Foundation For A Creative America.

Douglas E. Evelyn
Deputy Director
National Museum of American History

CONTENTS

Drawing from the catalog of the Chicago Model Works, manufacturer of small machinery, law court, and Patent Office models

OVERVIEW

The original seed of this catalog was "Patent Pending: Models of Invention," a 1986 exhibition in the National Museum of American History commemorating the 150th anniversary of the Patent Act of 1836. Following this exhibition, we were presented the opportunity of taking the show to Japan. As it was meant for an American audience, we had to transform it into one that would have meaning for the Japanese people. To do this, we produced a catalog in Japanese. The Japanese exhibition, like this catalog, addressed the interrelated areas of work and home. The main themes were Transportation, Domestic Life, Communications, Power and Energy, Agriculture, Mechanization, and Recreation. From the National Museum of American History's collection of 10,000 patent models, 113 were selected to illustrate the eagerness of nineteenth-century Americans to capitalize on new ideas, to build new machines, and to mechanize American life from factory to home. The exhibition also included 17 objects that were not models in order to provide a sense of realistic scale. Graphics depicted life in nineteenth-century America, the U.S Patent Office, and individual inventors and their inventions.

Now we are pleased to present an English-language version of the catalog in commemoration of the 200th anniversary of the Patent Act of 1790. To augment this celebration, we produced "Icons of Invention," a smaller exhibition of patent models, at the National Portrait Gallery. The site was especially appropriate since the National Portrait Gallery occupies the former U.S. Patent Office, the original home of these models in Washington, D.C. Across the Potomac in Crystal City, Virginia, the present site of the Patent and Trademark Office, two other exhibitions celebrated the patent and and copyright bicentennial. They were "Her Works Praise Her," an exhibition of inventions by women, and "Recognizing Minority Inventive Genius," an exhibition of inventions by African, Asian, Hispanic, and Native Americans, Pacific Islanders, and Alaskan Natives. The Smithsonian's Anacostia Museum also featured an exhibition, "The Real McCoy: African American Inventors and Innovations."

These exhibitions bring to mind a familiar American myth: The poor inventor, with hard work and inspiration, becomes rich and famous. In reality, few succeeded in the 1800s. Most patented inventions were neither important nor commercially successful. Many patents issued in the nineteenth century were for minor improvements to an existing device rather than for wholly new or original technologies. Although individual inventions capture our imagination, it is important to see them as components of larger and more complex systems. Thomas Edison's light bulb, for example, only had value as part of an electrical system.

Each country writes its own history with the nuances of its language and the shared history of its people forming the framework of understanding. By exhibiting these patent models and describing the context of their invention, we hope to impart an understanding of the development of technology and industry in America during the nineteenth century. Visually, these models remind us of the men and women who dreamed of new approaches and searched for solutions to problems, hoping to make a difference by patenting their ideas.

Barbara Suit Janssen
Project Director And Exhibition Curator
National Museum of American History

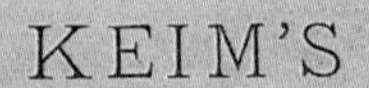

KEIM'S

ILLUSTRATED GUIDE

TO THE

MUSEUM OF MODELS,

PATENT OFFICE.

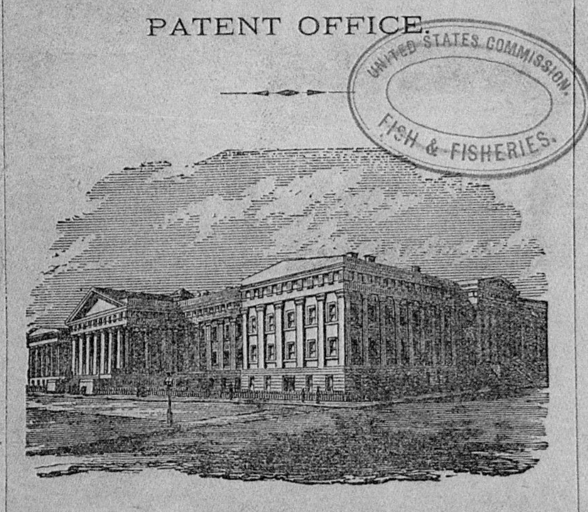

WASHINGTON, D. C.

DeB. RANDOLPH KEIM.

1874.

Guide Booklet for visitors to the Patent Office's Museum of Models

★ Patenting and Patent Models in Nineteenth-Century America

Although the establishment of a national patent system in 1790 was one of the first acts of the new United States government, the practice of patenting inventions already had a long history in America. Before 1790, American patent practice was based on English patent law—as was the first national patent statute itself. But like the earlier practice, the new statute also contained some peculiarly American features. Among these was the requirement that applicants for patents submit models of their inventions to the patent-granting authorities whenever an invention could be represented by a model. English law required drawings and a written description of the invention, as did the new American law, but the American law was the first to require a model as well. There were reasons for this unusual stipulation.

Before 1790, state and colonial authorities often granted patents or patent-like monopolies for inventions which subsequently turned out to be no more than impractical dreams. To protect the public from such useless and cumbersome monopolies, these authorities eventually began to require models from petitioning inventors as evidence of the operability of their inventions. In other instances, they required patent applicants to deposit models of their inventions in an important city where they could be freely inspected by the public. The authorities saw this as a way of insuring that others could learn to make or use the invention after the patent expired, or in case an inventor failed to make his invention available to the public before then. Finally, since the officials called upon to grant patents often lacked the technical background necessary to understand a complex invention solely from a written description or drawings, a model of the invention was often useful in helping determine its novelty and utility. In short, prior to 1790, models of inventions served two basic purposes: to demonstrate the operability of the invention, and to disclose it fully and accurately to government officials and the public. Those who drafted the new national law anticipated that models would be needed for the same reasons after 1790.

For three years the models submitted by inventors did indeed serve these purposes. In 1793 the initial statute was repealed and another quite different one took its place. In the past, patent applications had always been examined before patents were granted on them, but the new law called for no examination at all. Instead, it established a registration system. Patents were automatically granted to all applicants, and decisions regarding the patentability of the inventions covered by them were deferred until the patent was challenged in a court of law. Although inventors were still required to deposit models in the Patent Office, these were now used by the judges and juries rather than government officials. Like those officials, judges and juries usually lacked the technical background necessary to understand descriptions and drawings of an invention without the aid of a three-dimensional representation.

As a representation of the invention, the model now proved even more important than under the former law. Since the description and drawings submitted to the Patent Office were not examined beforehand, they were often found in court to be so incomplete and inaccurate that the invention could not be understood at all. This left the model as the only official evidence of what was intended to be covered by the patent. It was not unusual for the description of an invention in a patent to be completely rewritten, after the original version was found to be defective, with the Patent Office model serving as the only link between the two versions.

As proof of the operability of inventions, however, the importance of Patent Office model virtually disappeared. Since there was no longer any examination of the invention before a patent was granted, the model was no longer needed to demonstrate operability to Patent Office officials. On the other hand, if the question of operability arose in court, the inventor was free to prove it not just by a demonstration of the patent model but by any evidence he chose.

After forty years of experience with the registration system, it became clear to everyone that the practice of leaving the question of patentability entirely to the courts was causing a great deal of confusion regarding the validity of patents. Often, different courts reached different conclusions regarding the same patent, and no clear guidelines existed to ascertain whether a patent would be held valid or not. The system also proved to be particularly susceptible to fraud since it was very easy to copy a model on public display in the Patent Office, receive a patent on it, and then use it to extort royalties from an unsuspecting public. Pressure mounted to revise the patent system yet again, and the law was reformulated in 1836. The new law required an examination of an invention before a patent could be granted for it, but this time examinations were to be carried out by a corps of specially qualified examiners.

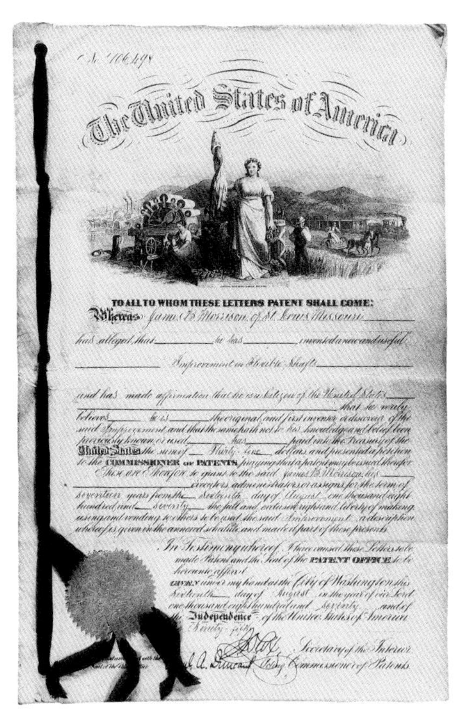
The United States of America

TO ALL TO WHOM THESE LETTERS PATENT SHALL COME:

Whereas James B. Morrison of St. Louis Missouri has alleged that he has invented a new and useful Improvement in Flexible Shafts and has made affirmation that he is a Citizen of the United States that he verily believes he is the original and first inventor or discoverer of the said Improvement and that the same hath not to his knowledge and belief been previously known or used has paid into the Treasury of the United States the sum of Thirty-five dollars and presented a petition to the COMMISSIONER OF PATENTS praying that a patent may be issued therefor.

These are Therefore to grant to the said James B. Morrison his executors, administrators or assigns for the term of seventeen years from the Sixteenth day of August one thousand eight hundred and seventy the full and exclusive right and liberty of making, using and vending to others to be used, the said Improvement a description whereof is given in the annexed schedule, and made a part of these presents.

In Testimony whereof I have caused these Letters to be made Patent and the Seal of the PATENT OFFICE to be hereunto affixed.

GIVEN under my hand at the City of Washington this Sixteenth day of August in the year of our Lord one thousand eight hundred and seventy and of the Independence of the United States of America Ninety-fifth

Secretary of the Interior

A. Duncan Acting Commissioner of Patents

James B. Morrison's patent papers for flexible shafts. Patent No. 106498, August 16, 1870, St Louis, Missouri

South Hall of Patent Model Room, United States Patent Office, 1888

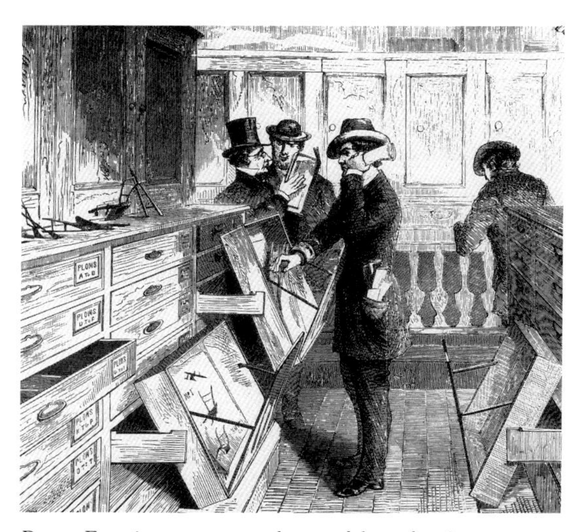

Patent Examiners compare plow models to drawings.
Harper's Weekly, July 10, 1869

In the reorganized patent system, patent models lost some functions and acquired others. Because the new examiners were well versed in the technology of the inventions they reviewed, they were able to understand an invention solely from its description in the patent application and its accompanying drawings. If these were found insufficient, the examiner could require the inventor to correct their deficiencies by amendment. Moreover, the examiners were usually able to determine solely from the description and drawings whether or not an invention was operable. Unlike the patent-granting officials under the 1790 examination system, these officials no longer needed the models to understand the inventions or to prove their operability. Nevertheless, the importance of the models as a medium of communication grew in a variety of ways.

Since a complete and accurate description of the invention, with references to accompanying drawings, was now a prerequisite to the granting of a patent, the model served, in the first place, as a medium of communication between the patent applicant, his draftsman, and his attorney. The draftsman rendered the model in the drawings to be sent to the Patent Office, and the attorney prepared the written description based on the drawings. One of the responsibilities of the patent examiner was to make certain that the model, the written description, and the drawings all agreed in their respective disclosures of an invention; basing both the drawings and the written description on the model helped applicants satisfy this requirement.

The models also served as a medium of communication within the Patent Office itself. Although they became less important to the examiners as a source of information about the inventions described in patent applications, they became more important as sources of information about what had already been patented. In 1790, rejections of patent applications for lack of novelty were usually based on the patent official's personal recollection of the technical literature. After 1836, however, as the number of patent applications and patented inventions both increased, it became more expedient to determine the novelty of new inventions by searching for similar devices amongst the models of patented inventions in the Patent Office. When such a search suggested possibly relevant patents, the examiner could consult the corresponding patent descriptions and drawings for additional details.

In much the same way, the model also served as an important medium of communication between the Patent Office and the public. Before applying for a patent, inventors, their attorneys, or patent agents often visited the model galleries of the Patent Office to determine whether something was indeed novel or whether it was already embodied in the model of some patent invention. To many other visitors, however, the models served as a medium of communication not for technical information but for something about America. Collectively, the models reflected the variety and profusion of American invention. In 1836 a fire destroyed the Patent Office building and all the models in it; plans for a new building, based on the assumption that the model collection would eventually serve as a monument to American genius and industry, called for a design appropriate to that function. And the splendid new Patent Office building became one of the capital's chief tourist attractions.

These various functions of the models are reflected in their design. They were usually built by the inventor himself or by a professional model maker working from a sketch or description furnished by the inventor. Since the model no longer had to prove the results claimed for the device by its inventor, it could be designed to emphasize whatever other information the designer chose. In some cases the models were perfect, scaled-down versions of the invention, but characteristically they exhibited obvious distortions in the sizes and shapes of their parts relative to their full-sized versions. In others, the distortions were simply the result of the need to make the model structurally sound. Many times, however, distortions were deliberately introduced to emphasize the model's novel features. When the invention was a simple improvement on an available device, what was old was depicted by the actual, improved device and what was new was simply added to it. Occasionally the old part was represented by a child's toy. These kinds of models are of particular interest today because they communicate something to us of their own history. Similarly significant are those models that are the actual experimental prototypes used by an inventor to perfect his invention. In these cases the model is often able to tell us something about the steps taken by the inventor in that process. In still other instances, the model designer chose to illustrate the new device in cross-section for better clarity; such models often were essentially two-dimensional, reflecting their function as draftsmen's models.

Because inventors knew that their models would be seen by thousands of visitors to the Patent Office, many chose to put extra effort into making them pleasing and attractive. These models therefore represent the attempts of their designers to communicate a typically nineteenth-century sense of the aesthetic quality of technology. Other inventors attempted to take commercial advantage of the popularity of the Patent Office as a tourist attraction by using their models as advertisements, their names and addresses prominently displayed in hope that they would be recognized in places where their inventions were offered for sale.

By 1870, the expense of maintaining the Patent Office model collection was becoming alarming. Their number had grown from virtually zero after the fire of 1836 to the hundreds of thousands; by all indications it was going to continue to grow. At the same time, many aging models were broken or so worn that they were useless as information sources. Gradually the government began to phase out the models as a primary source of patent information. In the early 1870s it began printing and distributing copies of all new patents as they were granted and all patents already in the files. The new printed patents, of a more convenient size and more readable than the handwritten copies they superseded, were readily accepted as substitutes by all those who had formerly relied on the models. By the end of 1880, the Patent Office no longer required the submission of models; in fact, except in certain cases, it refused to accept them. The models remaining in the Patent Office building were removed from public display and were eventually sold or otherwise disposed of.

Accustomed as we are today to printed patent documents, models seem a highly inefficient way to communicate information about patented inventions. But at some time in the not-too-distant future, when computers have become even more commonplace than they are now, our printed documents could seem just as archaic. The means we use to communicate information are like tools; we discard our old ones when better ones become available. Yet the old tools are worth saving for what they tell us about our past, and the old Patent Office models tell us a great deal.

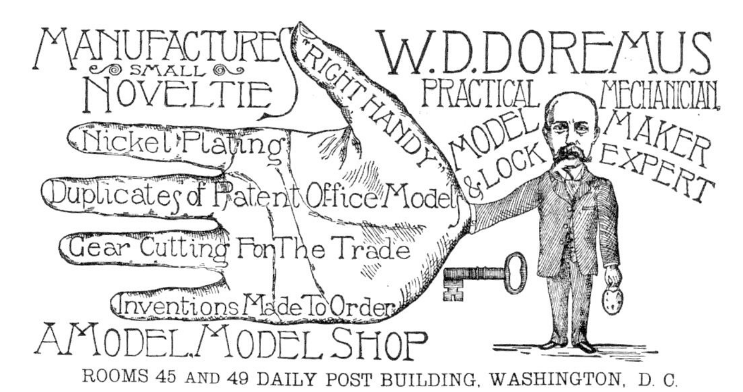

Patent model maker advertisement. *The Inventive Age*, December 30, 1890

Model maker in shop. *Harper's Weekly*, April 4, 1868

Patent Office fire, 1877

Patent drawing (signed LHL) of a Lamp Fixture for incandescent lamps designed by Lewis H. Latimer, a successful black inventor, while he was working for Hiram Maxim

★ Invention in Nineteenth-Century America

Because inventiveness became a mythic American quality, people have tended to assume that it was something apparent even at the time of independence from England in the 1780s. Actually it emerged rather slowly. True, there were some notable American inventors prior to the nineteenth century, among them Oliver Evans, John Fitch, Jacob Perkins, and, of course, Benjamin Franklin. Yet men like these often bemoaned the generally "low state of the mechanik arts." When Americans were forced to mechanize textile production after the Revolution, they did so by spiriting away sketches of British machinery. When Robert Fulton built the *Clermont*, the first practical steamboat in the United States, he ordered the engines from James Watt, an Englishman. When Evans sought to introduce mechanization to flour milling, he fought stubborn prejudice by millers. Charles Newbold's cast-iron plow of 1797 is said to have excited fears that it would "poison the soil." Nevertheless, the seeds of a fruitful inventive impulse had been planted in the early years of the American Republic, and they flowered when Americans were thrown back on their own resources as trade was disrupted during the Napoleonic Wars.

Within a generation, America was moving towards technological self-sufficiency. After the close of the second war with England (1812-15), attention focused on the nation's transportation needs. As the nationalist leader John C. Calhoun put it in 1817, "We are greatly and rapidly—I was about to say fearfully—growing. This is our pride and our danger; our weakness and out strength . . . . Let us, then, bind the Republic together with a perfect system of roads and canals." That same year, the New York legislature authorized construction of a canal connecting Lake Ontario at Buffalo with the Hudson River at Albany—linking the Great Lakes to the Atlantic Ocean. By the time this 364-mile Erie Canal was completed in 1825, however, talk of another mode of transportation was filling the air, and within five years the first American railroads were in operation. All the locomotives were initially imported from England. By the late 1830s, however, locomotives made in the United States were being sold abroad in Austria, Prussia, and England.

U.S. industrial exports rose throughout the 1840s, and by 1851, American goods—especially agricultural machinery, machine tools, clocks, locks, and repeating firearms—commanded a world stage at the Crystal Palace Exhibition in London, the "Exhi-

Artists and photographers record their excursion on the B&O Railroad, 1858, cameras in foreground.

Winans "Crab" locomotive, the *Mazeppa*, 1838, based on Winans's 1837 patents

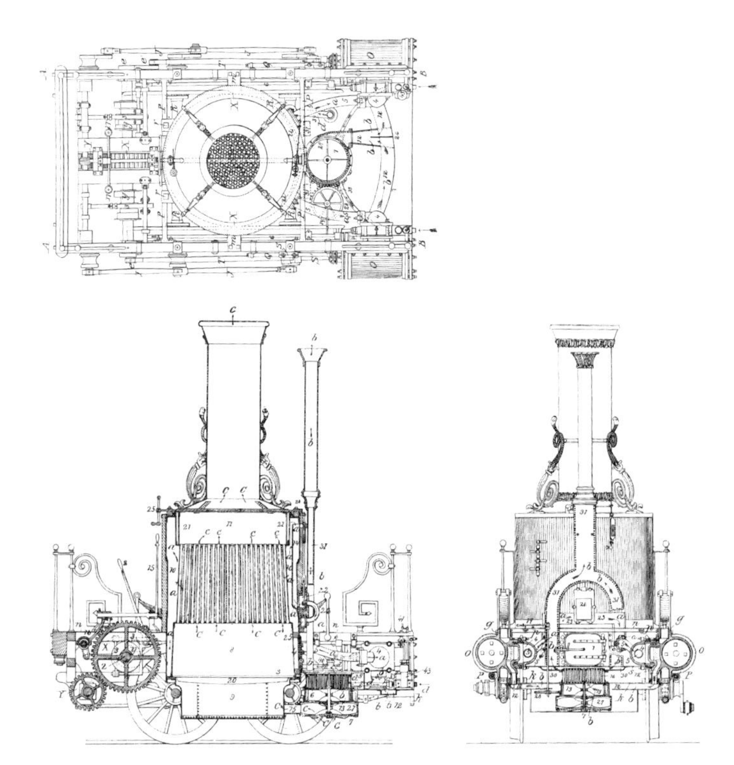

Locomotive patent drawings from Winans Patent No. 308, 1837

bition of the Industry of All Nations." Two years later a second Crystal Palace Exhibition opened, this time in New York City. Its promoters promised to showcase "the choicest products of the Luxury of the Old World and the Most Cunning Devices of the Ingenuity of the New." Thus the ground was prepared for emphasizing technological attainments in the American exhibits. In the realm of "cunning devices," a rich new orchestration unfolded, sufficiently rich to suggest that the New York Crystal Palace heralded the moment in American history when technology modulated from a minor to a major key.

The Crystal Palace exhibition introduced more American inventions by far than any previous exhibition in the United States. Nothing like it could have been marshaled ten years earlier, or even five. The most extensive exhibits were staged by the makers of agricultural implements, carriages and wagons, railroad equipment, and woodworking machinery. There were also displays by builders of ships and bridges; by makers of pumps, stationary engines, and mobile fire engines; rock drills and stone-dressing machines; and printing presses and machinery for casting and setting type. There were exhibits of punch, press, and shear machinery, sugar machinery, smut machinery, flax machinery, gins, looms, spinning frames, and tobacco presses. There were scales, meters, guns, locks, lamps, sewing machines, telegraph apparatus, and laboratory equipment; there was a shoe-pegging machine, a rag picker, an ice cream machine, a gold washer, a nail machine, an electric motor, and a brick machine. All told, there were at least one hundred exhibits of agricultural machinery, especially threshers, harvesters, mowers, and reapers. Of the latter there were ten in all, each patented; easily the most prominent was Cyrus McCormick's—the machine that, as one observer put it, "opened the eyes of our excellent neighbour, John Bull, to the genius and energy of Yankee farmers."

What the Americans exhibited in 1853 suggests a concern with speedy modes of transportation and communication, with efficient methods of exploiting resources, with keeping accurate track of time, measure, and quantity, and with mechanizing the production of consumer goods and the whole range of operations related to farming, food and textile production, and clothing manufacture. Contemporary Patent Office reports and comments by foreign observers such as Joseph Whitworth, the English toolmaker, confirm that these were precisely the realms of greatest inventive activity in mid-nineteenth-century America. True, there was no dramatic debut of whole new worlds of invention, nothing comparable to the acres of precision metalworking machinery at the Centennial Exposition in Philadelphia in 1876, or to the practical applications of electricity unveiled at the Columbian Exposition in Chicago in 1893. Some of the inventions exhibited at the Crystal Palace—the cotton gin shown by Eli Whitney's son, for instance—were already ensconced in the annals of American folklore. McCormick had built his first reaper, and John Stephenson, his first omnibus, some two decades before. The Woodworth wood planer and the steam-power inventions of George Corliss had become household terms as the result of patent litigation. It had been almost a decade since Samuel F.B. Morse inaugurated his commercial telegraph system, and by 1853 his company alone was operating more than 13,000 miles of line. Elias Howe's first sewing machine patent dated from 1846. Charles Goodyear discovered vulcanization in 1839. Samuel Colt was already a past master at publicizing his inventions. And so one might continue.

But the real element of novelty lay in the big picture: The concept of *mechanizing* all sorts of things from sewing to shooting, and the concept of mechanizing their *mode of manufacture*. Clearly such concepts were more congenial in the northern states than the South, for there were very few technological displays entered from any of those states that would secede from the Union eight years hence. New York City inventors predominated: Morse and Stephenson were headquartered there, as were most shipbuilders and instrument makers. Isaac Singer's was a New York City firm. Yet Elias Howe's sewing machine plant was in Cambridgeport, Massachusetts, Wheeler and Wilson's was in Watertown, Connecticut, and exhibitors of unpatented sewing machines came from such faraway places as St. Louis, Missouri, and Conneaut, Ohio.

Linus Yale, Jr., made his "Patent Magic Locks" in Newport, New Hampshire, Charles Goodyear made his India rubber goods in New Haven, Connecticut, William Jeffers made his fire engines in Pawtucket, Rhode Island, Junius Judson made his steam-engine governors in Rochester, New York, and E.H. Ashcroft made his steam gauges in Boston. McCormick's farm machinery factory was located in Chicago, Illinois. Fairbanks scales came from St. Johnsbury, Vermont. Hickcock ruling machines came from eastern Pennsylvania, Gallahue shoe-pegging machines from western Pennsylvania. The bridge builders Albert Fink and Wendel Bollman were located in Baltimore, Maryland. The Dodge brothers made their "Patent Premium Suction and Force Pumps" in Newburgh, New York.

As for woodworking machinery, the makers were spread all over the nation's northeastern quadrant. An American forte since the invention of Thomas Blanchard's copying lathe, special-purpose woodworking machines were found everywhere. On the other hand, only a few of the American metalworking machinery firms that were most prominent at the Philadelphia Centennial of 1876 exhibited in 1853, firms such as Pratt & Whitney, Browne & Sharpe, and William Sellers. Many were not even in business yet. Of the dozen or so nineteenth-century inventors who have passed from the history of American technology to its hagiography, only half were represented at the Crystal Palace. Thomas Edison, Alexander Graham Bell, and George Westinghouse were only children at the time, the others not yet born. The widespread introduction of such crucial machine tools as the Lincoln miller, the turret lathe, and the Spencer automatic screw machine was yet to come. The industries based on steel, petroleum, and electric power did not exist even in infancy. The most indelible image of the 1876 Centennial Exposition was the mammoth Corliss Engine, towering forty feet over the transept in Machinery Hall and providing motive power for hundreds of different factory machines. There is simply no analogous image for the Crystal Palace.

Still, the Crystal Palace embodied clear intimations of an awakening giant. One senses that Americans had definitely discovered their own inventive bent, that they knew—and now knew they knew—how to make things well, and make lots of them alike, and mechanize just about anything "assigned in the old work to manual skill." And all at once they could perceive the incredible multiformity of inventions to patent, to publicize, and to reap profit from. Six years before, the U.S. Patent Office had managed with a staff of sixteen; six years later it had seventy-five. The week that President Franklin Pierce inaugurated the Crystal Palace, sixteen U.S. patents were granted. Although the cumulative count

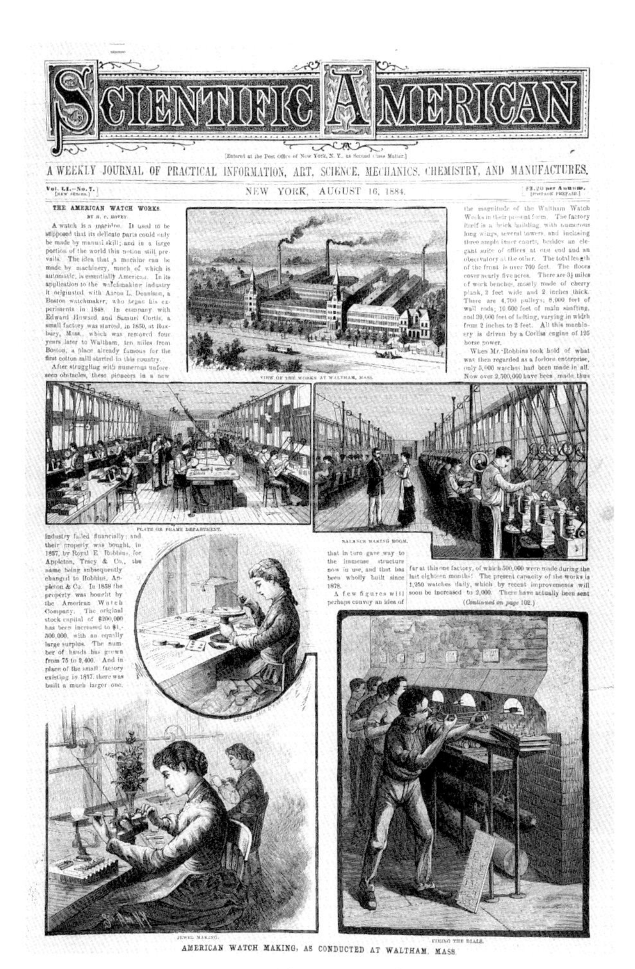
SCIENTIFIC AMERICAN

A WEEKLY JOURNAL OF PRACTICAL INFORMATION, ART, SCIENCE, MECHANICS, CHEMISTRY, AND MANUFACTURES.

NEW YORK, AUGUST 16, 1884.

$3.20 per Annum. [POSTAGE PREPAID.]

THE AMERICAN WATCH WORKS.

A watch is a machine. It used to be supposed that its delicate parts could only be made by manual skill; and in a large portion of the world this notion still prevails. The idea that a machine can be made by machinery, much of which is automatic, is essentially American. In its application to the watchmaker industry it originated with Aaron L. Dennison, a Boston watchmaker, who began his experiments in 1848. In company with Edward Howard and Samuel Curtis, a small factory was started, in 1850, at Roxbury, Mass., which was removed four years later to Waltham, ten miles from Boston, a place already famous for the first cotton mill started in this country. After struggling with numerous unforeseen obstacles, these pioneers in a new

industry failed financially; and their property was bought, in 1857, by Royal E. Robbins, for Appleton, Tracy & Co., the name being subsequently changed to Robbins, Appleton & Co. In 1858 the property was bought by the American Watch Company. The original stock capital of $200,000 has been increased to $1,500,000, with an equally large surplus. The number of hands has grown from 75 to 2,400. And in place of the small factory existing in 1857, there was built a much larger one, that in turn gave way to the immense structure now in use, and that has been wholly built since 1878.

A few figures will perhaps convey an idea of the magnitude of the Waltham Watch Works in their present form. The factory itself is a brick building, with numerous long wings, several towers, and inclosing three ample inner courts, besides an elegant suite of offices at one end and an observatory at the other. The total length of the front is over 700 feet. The floors cover nearly five acres. There are 3½ miles of work benches, mostly made of cherry plank, 2 feet wide and 2 inches thick. There are 4,700 pulleys; 8,000 feet of wall rods; 10,600 feet of main shafting, and 30,000 feet of belting, varying in width from 2 inches to 2 feet. All this machinery is driven by a Corliss engine of 125 horse power.

When Mr. Robbins took hold of what was then regarded as a forlorn enterprise, only 5,000 watches had been made in all. Now over 2,500,000 have been made thus far at this one factory, of which 500,000 were made during the last eighteen months! The present capacity of the works is 1,250 watches daily, which by recent improvements will soon be increased to 2,000. There have actually been sent

(Continued on page 102.)

American Watch Works, Waltham, Massachusetts. *Scientific American*, August 16, 1884

in the entire history of the American Republic had not reached 20,000, half of these dated from the seventeen years since the system had been reformed in 1836, and more than 1,000 from 1852 alone.

A steep upward trend was clear. The Civil War of the 1860s retarded it temporarily, but by the time of the Centennial Exposition, the Patent Office was routinely issuing more than 1,000 patents each month and the total had multiplied tenfold, topping 200,000. By the time of the Columbian Exposition in 1893, more U.S. patents were issued annually than in all the years prior to 1853 combined. While the curve would become even steeper in the twentieth century (U.S. Patent No. 1,000,000 was issued in 1911, No. 3,000,000 in 1961, and the number has now topped 4,500,000), it was obvious soon after the Civil War that America was emerging as the most innovative of nations, and would become the wealthiest as well as the premier exporter of manufactured goods—"peacefully working to conquer the world," in the phrase associated with the epitome of American entrepreneurial prowess, the Singer sewing machine.

This turn of events was not yet apparent at mid-century. Still, an acute observer could perceive the exponential nature of invention in America. In 1844—the year that marked America's "takeoff" into sustained economic growth—Charles Page, the Patent Office's Chief Examiner, observed in his annual report that

Men of Progress, painted by Christian Schussele, 1862. National Portrait Gallery, Smithsonian Institution; gift of Andrew W. Mellon

"Man's wants increase with his progress in knowledge; and hence the paradoxical truth, that the growing number of inventions, instead of filling the measure, increases the capacity." That the nation's economic takeoff coincided with this increase in capacity is not a happenstance. It is happenstance that it also coincided with a time when inventors were required to submit patent models. Yet this coincidence endows the models with a signal relevance to the history of invention and enterprise. They stand as a testament to an epoch which might be termed America's Mechanical Age.

To be sure, many species of models are not primarily mechanical, nor mechanical at all, and yet the majority of inventions represented by models fit somewhere under the rubric of mechanics. In the last two decades of the nineteenth century, after the model requirement was rescinded, inventions in the realms of applied electricity, chemical engineering, thermodynamics, and cybernetics signaled a qualitative shift as significant as the quantitative amplification at mid-century—a shift so important that it has been dubbed the "Second Industrial Revolution." Hence, patent models as representations that somehow embody a mechanical arrangement of "several parts" are by and large congruent with a discrete epoch in the history of American invention. As Kendall Dood writes, "the patent model requirement was uniquely suited both to the state of American technology in the mid-nineteenth century and to the popular conception of technology as *something entirely and immediately apprehendable through the senses*, and it is this lost view of technology which the surviving models most uniquely memorialize."

The patent models also collectively constitute an unmatched symbol of a phenomenon that was long a keynote of America's sense of national destiny, something called "Yankee ingenuity." An oft-recited observation in the late nineteenth century, one usually attributed to a European observer at the Centenntial Exposition of 1876, goes like this: "The American invents as the Italian paints and the Greek sculpted. It is genius." Unburdened by conservative traditions typical of many other peoples, Americans had indeed shown a unique flair for invention. As we near the end of the twentieth century, there is some evidence that this gift has waned. Whether this is a long-term trend, or merely a lull, remains to be seen in the twenty-first century.

America on the Move

Railroads and Locomotive Technology

In 1830 the United States was a large land mass desperately in need of an adequate transportation system. Travel by roads was slow and expensive because of the dependence on horse-drawn vehicles. The rivers did not necessarily flow where commerce did, and they froze in the winter and dried up in the summer. Canals proved very expensive to construct, particularly in hilly or mountainous territory, and like natural waterways, they were encumbered by freezing and droughts. The answer to the young nation's transportation needs was found in the steam railway. Railroads could move a large amount of traffic in all seasons and over almost any terrain. Because the track could sustain the great weight of a steam locomotive, mechanical power could be more effectively applied there than on water or roadways.

The steam railway was perfected in Great Britain between 1810 and 1830, but few other nations embraced the idea as enthusiastically as the United States. By 1835 a railway mania was underway in this country that was to produce some 2,300 miles of line by the end of the decade. Construction began in the most populous eastern states, with most lines designed to connect a major seaport, such as Boston or Charleston, with the largely undeveloped hinterland west of the Appalachians. All lines faced a major challenge, not just in overcoming the mountain barrier but in garnering the financial resources needed for such ambitious projects. Most were private corporations, but all required and sought government support in one form or another. In a few instances, like the Western Railroad (Massachusetts), the construction was almost entirely financed with state funds.

By the early 1850s several trunk lines had crossed the eastern mountain ranges and were tapping into the agricultural and timber land beyond. The eastern states were now crisscrossed with a dense network of iron tracks. Lines were being built west of the Mississippi River by 1852, but the Father-of-the-Waters was not bridged for another four years. Total mileage stood at 18,300 in 1855; by then it was possible to travel from New York to Chicago at a rate of speed unthinkable at the beginning of the railway age, when travel time was counted in days rather than hours. The investment in the rail network was approaching $1 billion and represented the largest single capital commitment of any American industry. Yet this sum was to grow enormously as more money was borrowed at home and abroad for railroad construction. Periodic financial panics slowed but could not stop growth.

In 1860 the nation's first big business carried 2,600 million ton-miles of freight and 1,900 million passenger-miles. There was talk of tying the West Coast, particularly California, to the rest of the union with iron rails. The traffic potential did not warrant such a project, but political needs—plus the desire to satisfy the nation's urge for "manifest destiny" by linking the Atlantic and the Pacific—made the "transcontinental" a top priority. Construction began in 1863 and was completed ahead of schedule, in part because of substantial federal aid, on May 10, 1869. Other major western lines filled in more slowly, and only as the former Mexican and British lands were settled and developed. By 1893, just as the frontier itself was declared "closed," the basic U.S. rail network was in place. Peak mileage was reached in 1916, with some 254,000 miles.

1.
Locomotive
Ross Winans
Baltimore, Maryland
July 29, 1837
Patent No. 305

After World War I automobiles and motorbuses began to erode railroad dominance in the intercity passenger business, while motor trucks began to take over the freight traffic, especially the short-haul trade. Improved roads accelerated the diversion of traffic away from the railroads. Commercial aviation began to cut into long-distance passenger travel markets after World War II, and today commercial passenger travel by rail is minuscule. Only about 35 percent of the nation's freight traffic goes by rail, much of this in low-rate bulk goods such as coal and grain. The railroads no longer dominate the nation's transportation business. Nevertheless, it appears that they will continue to play an important, if diminished, role in the economic affairs of the United States well into the twenty-first century.

Some inventors were convinced that making radical changes to the general arrangement of the steam locomotive was the only way to improve its performance. One of them was Ross Winans of Baltimore, whose designs were consistently at odds with standard practice. Winans produced a long line of oddities and curiosities which, for all of their novelty, were no more efficient than the machines they were intended to supersede.

The general design illustrated by this patent model was documented in five patents, all issued on July 29, 1837 (Nos. 305, 307, 308, 309, and 311). Winans modified the early "Grasshopper" locomotive by positioning the cylinders horizontally rather than vertically. A double countershaft with gears was attached to the driving axles through outside cranks and side rods. The advantage of such an expensive and complicated arrangement is difficult to understand. Winans also insisted on using vertical boilers, which only added to the machine's awkwardness and inefficiency. Only a few locomotives were built on this plan in 1837 and 1838. They became known as "Crabs" because of their peculiar appearance. Despite the obvious defects in their design, they continued in service until about 1870. Winans went on to develop an eight-wheel version of the Crab, of which about a dozen were produced.

This model is larger, more complete, and more carefully detailed than the average patent model. While most locomotive patent models are made to a scale of ½ inch to the foot (1:24) or smaller, this one calculates roughly to a scale of two inches to the foot (1:6). It seems ironic that such a handsome and elaborate model was made for a design of such modest potential.

This lithograph of a Hinkley 4-4-0 locomotive built in 1870 shows that basic locomotive design was *not* a matter of patented features. (See page 74) The general arrangement included the following elements: a horizontal firetube boiler, level cylinders, a direct connection from the pistons to the wheels by rods (i.e., no gears or belts), and a truck usually made up of a pair of smaller wheels which was

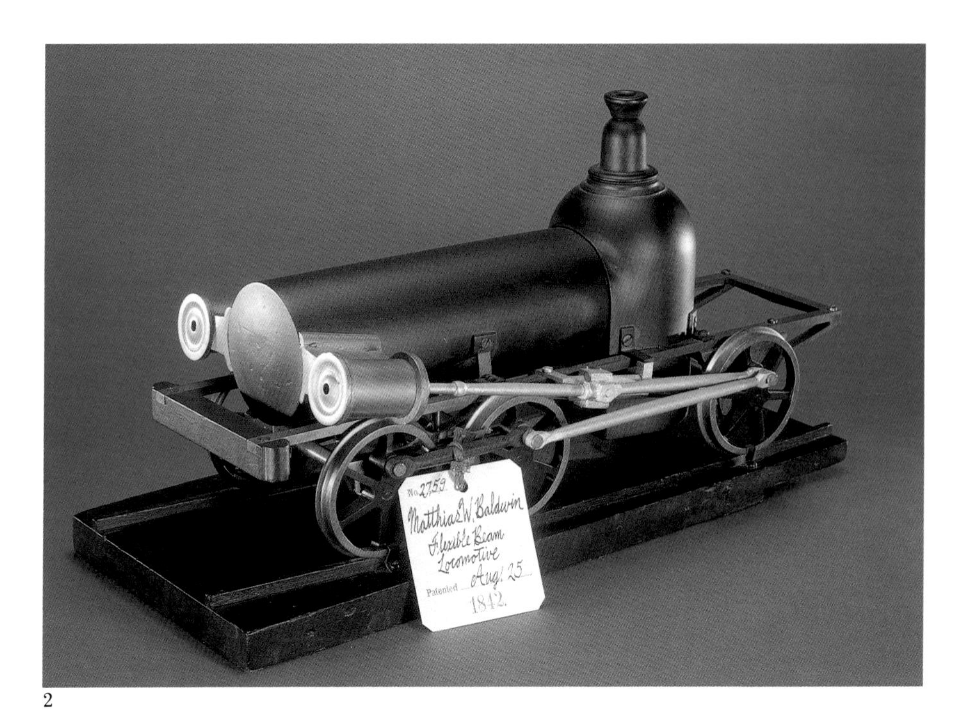

2

3

2.
Locomotive
Matthias W. Baldwin
Philadelphia, Pennsylvania
August 25, 1842
Patent No. 2759

3.
Locomotive
Andrew Cathcart
Madison, Indiana
October 23, 1849
Patent No. 6818

meant to guide or steer the engine, especially on curves. Not one of these features was patented. Nor were minor generic features such as the pilot or cowcatcher, the bell, the cab, and the whistle. Patented bells and whistles did exist, but they were of peculiar design and often as not were never used.

The general features shown here evolved rather early in the history of the steam locomotive; in fact, the basic shape and plan was fixed in England by Robert Stephenson in 1829 and 1830. Most of these features were either obvious or were borrowed from others active in the field. The firetube boiler, for example, was supposedly suggested by one of Stephenson's associates but the scheme actually dates back to Roman times. The leading truck was introduced in 1831 by an American engineer, John B. Jervis, and yet even its originality is not entirely unclouded, for a British mechanic named Chapman had suggested a similar plan some nineteen years before. Jervis understood the value of his idea but never patented it. And so it could be argued that one of our most useful mechanical servants, the steam locomotive, came into full flower essentially in isolation from the patent system.

M.W. Baldwin of Philadelphia was one of America's earliest and most productive locomotive builders. Like others in this trade, he sought ways to build engines better suited to the rough and uneven tracks of early American railroads. Special attention was given to devising ways to make the engine handle curves with greater speed and safety. In his "flexible beam" plan, Baldwin added a secondary frame to the two front pairs of driving wheels; this secondary frame was formed from two beams held in place by a ball joint and thus able to turn and tilt freely. The axle bearings were conical so they could turn freely as well. The axles could slide left or right of the bearings. In all, the running gear was extremely flexible.

Baldwin patented his plan on August 25, 1842 (No. 2,759) and built the first engine on this general design, the *Tennessee* for the Georgia Railroad, in December of that year. The first flexible-beam engines were of the six-wheel pattern but in 1846 eight-wheelers were offered as well. About three hundred such engines were built between 1842 and 1866. During these same years Baldwin built some 1,350 locomotives in all, indicating that the flexible beams were not among his most popular.

Like most patent models, this one is not a working model. Stripped down to the bare essentials and such important elements as the smokestack, valve gear, and feedwater pumps, it merely demonstrates the principles of the invention.

Overcoming changes in elevation is a fundamental problem for all railroad operations. Conventional locomotives are used by most railroads because grades are usually held to one or two percent. Mountain scenic railways such as those on Mount Washington or Pikes Peak require rack-and-pinion arrangements: a tooth or rack rail placed in the center of the track is engaged by a gear fixed to the locomotive's driving machinery. Rack locomotives are geared down for power and creep up the steeply inclined track very slowly.

The only example of a main-line railroad using the rack system was found on the Madison & Indianapolis, a pioneer Midwestern railroad that opened in 1838. A short section of the line, from the Ohio River to the top of the hills at Madison, Indiana, was originally worked by horses, one car at a time. In 1847 Andrew Cathcart, master mechanic of the M&I, developed a plan for a rack locomotive which featured a separate drive for the spur gear. On level tracks, the engine operated as a conventional adhesion locomotive; when on the steep grade, the second pair of cylinders, placed vertically astride the boiler, worked a separate crank axle that powered the center rail gear.

Cathcart patented his invention on October 23, 1849 (No. 6,818). He lived to see one more engine built on this plan, but his system was never used anywhere except on the Madison grade. There it worked moderately well, though it proved both dangerous and expensive to maintain. In July 1868 the Cathcart system was abandoned in favor of very heavy tank engines that could overcome the 5.8 percent grade without rack and pinion.

The Selden Automobile Patent

George Selden's dubious claim that he invented the combustion-engine automobile ran through the young American automobile industry like a virus. His claim rested on an application for a patent on a "road-locomotive" that he had filed in 1879; he deliberately delayed approval of the patent until 1895, however, when automobiles were attracting the attention of the American public, inventors, and

4.
Automobile
George B. Selden
Rochester, New York
November 5, 1895
Patent No. 549160

manufacturers. By 1912, when the patent expired, it had spawned a licensing association that exacted royalties from automobile manufacturers and served as a target for Henry Ford's much-publicized allegations of exploitation by automobile corporations. While the Selden patent itself had little influence on the design of the automobile, for seventeen years Selden's name ranked with those of leading automotive manufacturers, inventors, and engineers who built, modified, and sold automobiles in ever-increasing numbers.

George Baldwin Selden was born at Clarkson, New York, a village near Rochester, on September 14, 1846. He attended the University of Rochester and the Sheffield Scientific School at Yale University. He then studied law and was admitted to the bar in 1871. Selden was fascinated by mechanical problems, especially road locomotion, and he obtained a patent for a rubber tire for bicycles as well as patents for a typewriter and a machine for making barrel hoops. Professionally, he began to specialize in patent law.

In 1877, intrigued by the challenge of devising an engine light enough to propel a road vehicle, Selden designed a three-cylinder combustion engine with an external compression air pump opposing each cylinder. This proposed engine was a smaller, improved version of George Brayton's two-stroke, external-compression engine of 1874. In 1878 Selden commissioned a prototype of his engine with only one working cylinder.

Selden filed a patent application on May 8, 1879 for "any form of liquid-hydrocarbon engine of the compression type" combined with a chassis and broadly defined mechanical components, including front-wheel drive, a steering mechanism, clutch, and brake. The non-operating, metal model of the conceptual vehicle that he submitted is today in the collection of the National Museum of American History, Smithsonian Institution. During the 1880s Selden asked several individuals for money to construct a full-size, working vehicle according to his patent description, but no one would sup-

5.
Ford Model T Runabout
1926
Catalog No. 333777

port the project. He decided to delay approval of the patent by filing a series of amendments, often days before the two-year grace period expired. By this means he delayed issuance of the patent until November 5, 1895, at which time his proposed vehicle received Patent No. 549,160.

Although combustion-engine automobiles had been produced commercially in Europe since the early 1890s, and Selden had never built a working vehicle, some pioneer American automobile makers respected the Selden patent because it seemed to describe in basic terms the combination of a combustion engine and a chassis. In 1896 Alexander Winton of Cleveland, Ohio, paid Selden $25 for a ninety-day option to build automobiles. Three years later Hermann F. Cuntz, a mechanical engineer and patent expert for the Pope Manufacturing Company of Hartford, Connecticut, brought Selden's patent to the attention of William C. Whitney, financier of Pope's taxicab affiliate, the Columbia and Electric Vehicle Company. Late in 1899 the C&EVC paid Selden $10,000 for exclusive rights to administer the patent as protection in the event that the firm should begin manufacturing combustion-engine automobiles.

In June 1900 Whitney's attorneys notified several automobile manufacturers that they were infringing upon the Selden patent. In July the C&EVC's financially depleted successor, the Electric Vehicle Company, filed suit against the nation's leading manufacturer of combustion-engine automobiles, the Winton Motor Carriage Company of Cleveland, Ohio. The Winton forces first tried to have the suit dismissed on grounds that the Selden patent lacked validity. This tactic failed, however, and after two years of litigation the Winton company, faced with mounting legal costs and signs that other firms might capitulate to the Selden stranglehold, tried to negotiate a settlement out of court.

Meanwhile two automobile executives, Henry B. Joy of Packard and Frederic L. Smith of Olds, took the Selden matter into their own hands. They demanded that the automobile industry at large administer the Selden license, and they threatened to support the Winton defense if the Electric Vehicle Company would not cooperate with them. An agreement was reached, and the Association of Licensed Automobile Manufacturers (ALAM) was formed in March 1903. Manufacturers of combustion-engine automobiles were expected to pay the ALAM royalties amounting to 1.25 percent of their gross retail income. One-tenth of the money collected would go to Selden, one-tenth to George H. Day, general manager of the ALAM and president of the Electric Vehicle Company, two-fifths to the Electric Vehicle Company, and two-fifths to the ALAM for rebates to member firms and for a fund set up to finance infringement suits. The Winton company acknowledged the validity of the Selden patent, and the case against the pioneer auto maker was closed.

Many of the firms that joined the ALAM were manufacturers of high-price, low-volume luxury automobiles. The organization's membership policy, though vaguely formulated, tended to exclude small manufacturers, and licenses were difficult to obtain. The ALAM sought to discourage competition while protecting the flow of income to its members; when accused of creating a monopoly, its spokesmen claimed that the organization was simply helping to stabilize the industry by preventing overproduction and excessive cheapening of the product. The ALAM tried to limit entrance into the industry by suing firms that "borrowed" the technology that it purportedly controlled, by requiring previous automotive experience and integrated production before granting a license, and by prohibiting ALAM-affiliated dealers from selling automobiles that had not received a Selden license. These policies might have curtailed the development of quantity production of low-priced automobiles, which were in great demand among farmers and Midwesterners, had the ALAM been totally successful in its restrictive measures. As it turned out, the ALAM did not succeed in inhibiting the establishment of manufacturers in the low-price field, but it did try to intimidate consumers who considered buying unlicensed products. The slogan "Don't buy a lawsuit with your car" appeared in ALAM advertisements.

In 1903 the newly formed Ford Motor Company tried to obtain a Selden license but was turned down because Henry Ford's first two attempts to manufacture automobiles had been unsuccessful and because of the ALAM stance against competition in the low-price field. Ford first asked the ALAM to reconsider its decision, but he was so antagonized by Selden's self-designation as the father of the automobile industry and the ALAM's opposition to his attempt to mass-produce a low-priced automobile that he decided to attack the ALAM-Selden stranglehold. Ford and the ALAM exchanged opening blasts in newspapers and automotive trade journals. Then in October 1903 the Electric Vehicle Company and George Selden filed suit against the Ford Motor Company for infringement of the Selden patent. Observers of the automobile industry witnessed the opening pyrotechnics of a historic patent battle that reached its climax more than seven years later with a dramatic, decisive victory for Ford.

Henry Ford used the trial and the attendant publicity to cast himself as a pioneer automobile designer and individual entrepreneur locked in a David-and-Goliath struggle against powerful, oligarchic interests determined to stifle competition and raise prices for consumers. This message had great appeal in an era when price controls exercised by huge corporate trusts angered and frightened average citizens, an era when consumers and small entrepreneurs hoped for the restoration of free competition and fair play in the marketplace. The press sympathetically (and erroneously) referred to the ALAM as the "automobile trust," and Ford used the controversy to fix forever in the public's mind his image as a friend of the average consumer. The court victory raised Ford's stature in the automotive industry and established him as an independent thinker among corporate conspirators, a champion of reasonable and beneficial goals opposed by unreasonable but powerful special interests.

Ford's defense, led by attorney Ralzemond A. Parker, attempted to show that Selden's 1879 design for an engine and chassis was not original and that Selden could not possibly be credited with anything beyond a refinement of the Brayton engine. They showed that mechanical devices such as a clutch and steering mechanism had been used before 1879 and that an experimental motor omnibus had been fitted with a Brayton-type combustion engine in 1878. As further evidence that Selden was not the first person to make a combustion engine capable of propelling a chassis, Ford's shop workers constructed a working engine resembling a variation of Etienne Lenoir's 1860

6

7

8

9

6.
Columbia Bicycle
1886
Catalog No. 307217

7.
Tricycle
Otto Unzicker
Chicago, Illinois
June 4, 1878
Patent No. 204636

8.
Tricycle
Charles Hammelmann
Buffalo, New York
March 2, 1880
Patent No. 225010

9.
Tricycle
Francis Fowler
New Haven, Connecticut
February 3, 1880
Patent No. 224165

gas engine that Alexander H. Brandon patented in 1869, and they installed it in a 1903 Ford automobile chassis.

But Selden and the ALAM had their own arsenal of words and exhibits. Dugald Clerk, an international authority on combustion engines, argued that the Brayton engine did not differ significantly from Nicolaus Otto's four-stroke, internal-compression engine of 1876, the basis of nearly all combustion automobile engines then in production. He also argued that Selden's refinements to the Brayton engine represented a quantum leap in engine design and portability and made possible the modern automobile. During the trial two versions of the Selden vehicle were built more or less according to the 1879 patent design; one incorporated Selden's actual 1878 engine, modified for three-cylinder operation, but the other had features that were not even mentioned in the patent. In June 1907 both of the Selden-inspired vehicles were test-driven on a racetrack near Guttenberg, New Jersey, with poor results.

The verbal and mechanical demonstrations nevertheless impressed the judge hearing the case, Charles M. Hough, who had admitted that he did not possess a good understanding of mechanical devices. Hough waded through an obtuse and highly technical written statement by Ford's attorney, and a much simpler statement by the ALAM. He made his decision based on the superficial originality of Selden's "invention" and the automatic protection provided by patent law. Brushing aside pre-1879 engine patents and experimental vehicles, the United States Circuit Court of the Southern District of New York upheld the claim against Ford in

September 1909. It acknowledged that the Selden patent applied to any type of combustion engine combined with a road vehicle, and it recognized Selden as the man who had invented the modern automobile by conceiving of such a juxtaposition.

Since the Selden patent would be in effect for three more years, Ford took his case to the United States Circuit Court of Appeals. Fortunately, from Ford's point of view, one of the appellate judges, Walter C. Noyes, had a firm understanding of automotive mechanics and engines. Ford's attorneys restated their arguments, carefully comparing Selden's work with earlier inventions. An attorney for Panhard et Levassor, a co-defendant in the case, helped turn the tide by producing a recent book about combustion engines written by Dugald Clerk. The attorney pointed out that Clerk did not mention Selden in his book and credited others with developing successful combustion automobiles. He quoted from Clerk's book as follows: "No one, however, has yet succeeded in carrying Brayton's engine further than he [Brayton] did." This statement contradicted Clerk's earlier testimony lauding Selden's "improvements" to the Brayton engine.

In January 1911 the Court of Appeals unanimously ruled in favor of Ford, reversing the lower court's decision and limiting the Selden patent's validity to vehicles with Brayton-type engines as modified by Selden. Excluded from the scope of the Selden patent were currently produced combustion-engine automobiles, almost all of which had engines derived from the Otto engine. As a result of this ruling, royalty payments to the ALAM ceased immediately. The organization was dissolved in January 1912.

The ALAM, though short-lived, made one lasting contribution to the development of the automobile industry by promoting technical standards for materials, parts, processes, and tools used in automobile manufacturing. This work was carried out between 1903 and 1909 and helped to lay the groundwork for industry-wide standardization. In 1910 the ALAM's engineering library, records, and apparatus were transferred to the Society of Automobile Engineers (later renamed the Society of Automotive Engineers), which assumed an even greater role than the ALAM in the establishment of technical standards for the industry.

After his defeat in the judicial process, George Selden fell back on another enterprise: the Selden Motor Vehicle Company, which he had established in Rochester, New York in 1906. Until 1914 this firm built automobiles with conventional engines, and beginning in 1913 it built motor trucks.

For the rest of his life, Selden privately clung to his belief that he was the father of the modern automobile. Yet at the time of Selden's death, which occurred on January 17, 1922, Henry Ford stood alone as the "father" of low-priced, mass-produced automobiles for average Americans, and as the nation's largest producer of automobiles. Ford held no bitterness toward his former nemesis, but his policy of inclusiveness clearly had triumphed over Selden's vision of exclusiveness.

Ford Model T Runabout

The Ford Model T was the first automobile owned by large numbers of farmers, workers, and other middle-income Americans. Well suited to farm chores and country roads, the Model T, like Henry Ford himself, struck a responsive note among rural American who wanted practical, low-cost automobiles. Only a small number of farmers could afford the first Model T built in 1908, but by 1915 the Ford moving assembly line cut prices in half. By the early 1920s the Model T was by far the best-selling and least-expensive new car, outselling all other makes combined. The Ford Motor Company produced 17,771 Model T automobiles in 1909. In 1915 production jumped to 501,462 and in 1923 escalated to 1,817,891. Model T ownership brought independent mobility to millions of rural and urban Americans, but in its last two years of production, 1926 and 1927, the Model T was eclipsed by more luxurious General Motors automobiles and by inexpensive used cars.

The Model T resembled both the early gasoline buggies and contemporary engine-in-front automobiles. It sat high over the road so that hardened earth and dust were less likely to damage the engine and chassis. A three-point suspension system enabled the car to travel with ease on bumpy roads. The five-passenger touring car provided more seating space than the runabout, but both body styles were popular. The four-cylinder, twenty-horsepower engine was powerful relative to the light body.

Bicycles and Tricycles

Riding high-wheel bicycles became a popular form of touring and outdoor exercise in the United States after 1876, the year an English high-wheeler was first imported. The high front wheel and saddle provided speed and stability but caused numerous injuries and prevented women from riding. Large tricycles, also developed in England, allowed women to go cycling, but were heavier than bicycles and harder to pedal. During the 1880s standard tricycles and high-wheel bicycles were manufactured in large quantities in the United States. At the same time, variations of wheel arrangements, propulsion, and seating on tricycles and bicycles were proposed, patented, and sometimes built by American inventors who sought to make cycles safer and easier to ride. During the "bicycle craze" of 1878-1900, millions of American cyclists had their first encounter with personal, mechanized transportation. The popularity of the bicycle stimulated many innovations in mechanical design and mass production that the automobile industry later borrowed.

Otto Unzicker's tricycle, patented June 4, 1878 (No. 204,636), resembled a child's velocipede of the 1870s and it was designed to accommodate a female rider. Instead of foot pedals for propulsion, it had reciprocal, pivoted handlebars and cranks that both rotated and steered the front wheel. The side saddle and stirrup kept the rider's skirt off the frame and were patterned after ladies' riding gear.

The design of Francis Fowler's tricycle, patented February 3, 1880 (No. 224,165), with its large forward driving and steering wheels and curved backbone, resembled the high-wheel bicycle of the same period. A ratchet mechanism helped to maintain stability by allowing the outer wheel to revolve freely while the vehicle made a sharp turn; this mechanism was a forerunner of the modern automobile differential.

Charles Hammelmann's tricycle, patented March 2, 1880 (No. 225,010), differed from standard tricycles in its rack-and-ratchet drive mechanism connected to spring-returned foot levers. The vehicle's design also placed the rider almost directly over the front driving wheel.

Mississippi River steamboats wait for cargo in shoal water south of Memphis, Tennessee.

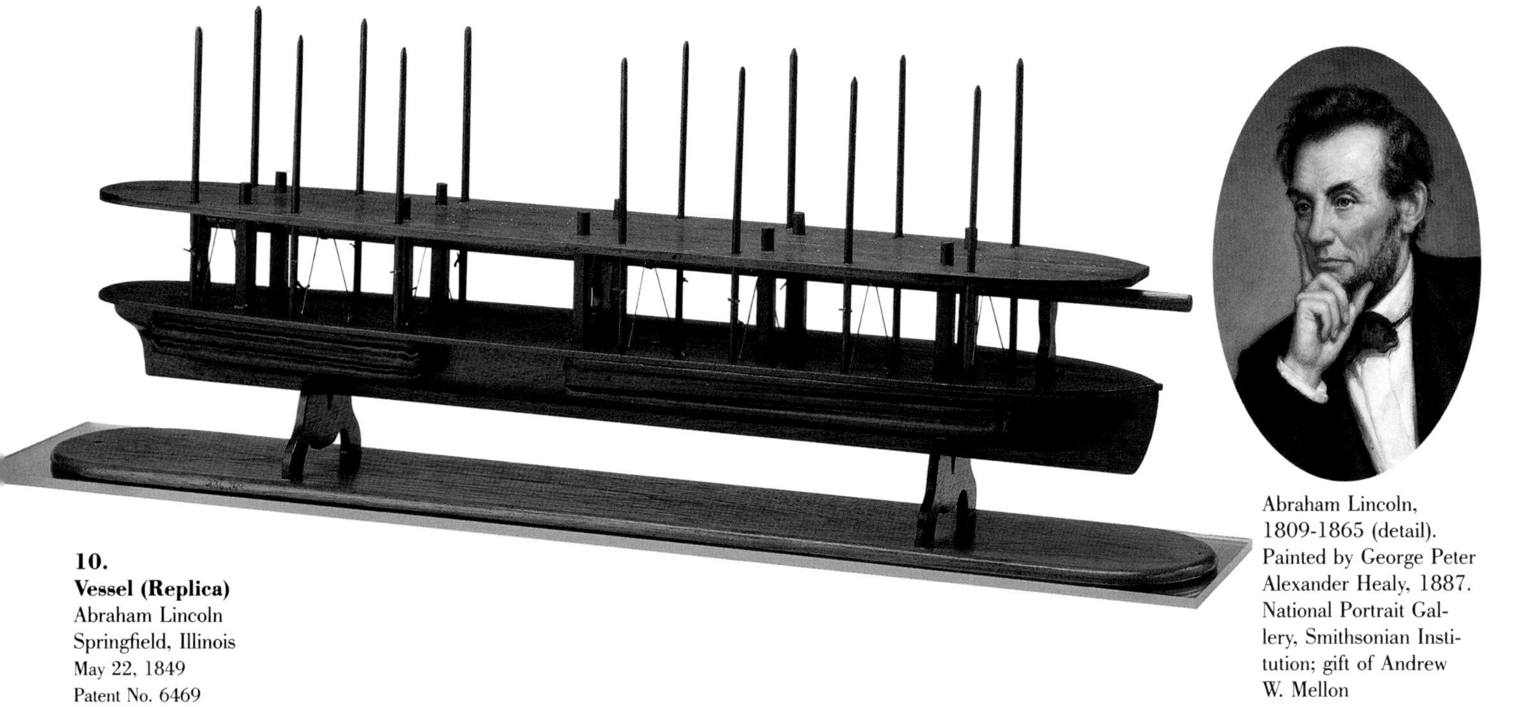

10.
Vessel (Replica)
Abraham Lincoln
Springfield, Illinois
May 22, 1849
Patent No. 6469

Abraham Lincoln, 1809-1865 (detail). Painted by George Peter Alexander Healy, 1887. National Portrait Gallery, Smithsonian Institution; gift of Andrew W. Mellon

11

12

13

11.
Steamboat Steering Gear
Frederick E. Sickels
New York, New York
May 10, 1853
Patent No. 9713

12.
Steam Steering Apparatus
Frederick E. Sickels
New York, New York
July 17, 1860
Patent No. 29200

13.
Paddle Wheel
Fletcher Felter
Perth Amboy, New Jersey
November 8, 1854
Patent No. 11992

ABRAHAM LINCOLN

As the only president who ever submitted a patent model, Lincoln took a personal interest in inventions. At age forty he obtained a patent for a device to raise steamboats off sandbars. His simple wooden model shows the patented features clearly. On both sides of the vessel are buoyant air chambers that expanded and contracted through attached vertical shafts sliding up and down. When the chambers were expanded, the additional air would float the vessel off sandbars and through shallow water. The chambers were to be made of waterproof cloth. Although Lincoln's flotation device was never manufactured, he remained interested in invention and encouraged the use of new technologies during the Civil War. Lincoln once observed that "the patent system added the fuel of interest to the fire of genius."

The *Jno. A. Scudder*, a steam-powered, side-wheel riverboat, carried cotton bales from Memphis to New Orleans on the Mississippi River from 1873 to 1898.

INSTANT COMMUNICATIONS

★ THE TELEGRAPH AND THE TELEPHONE

★ PRINTING

★ PHOTOGRAPHY

★ TYPEWRITING

EDISON TIN-FOIL PHONOGRAPH

None of Thomas Edison's inventions excited as much public interest as his speaking phonograph. Called "the miracle of the nineteenth century" by its promoters, the machine was the first ever to reproduce the sound of the human voice. Edison introduced the machine to the public in early 1878 but did not develop the machine for wide commercial distribution until competition arose about a decade later.

This machine is unmarked, but similar in appearance to those made in small quantity for Edison by Sigmund Bergmann of New York in about 1878. The phonograph operates on the same principles as Edison's earliest machines. In response to sound vibrations, a stylus attached to the reproducer makes indentations on a sheet of tin foil wrapped around the cylinder. Edison's later machines recorded sound on wax cylinders.

THE TELEGRAPH AND THE TELEPHONE

The first practical fruits of the new science of electricity were being felt in the 1830s, just as the Patent Office began to require that inventors furnish models when applying for patents. We therefore find among the models an especially good representation of the formative years of electrical technology. They cover all of those nineteenth-century innovations in current electricity that set the stage for an electrified world. The requirement for models ended, however, just before the new era of electronics. In the virgin territory covered by electrical technology, there was much room for innovation; it is not surprising that several who worked in this area were among the most prolific American inventors. Thomas Edison (1843-1931) still holds the record, at 1,096, for the largest number of patents granted, most of them electrical in nature.

The first significant electrical technology was the telegraph. The principles of electric telegraphy were conceived independently by several people in the late eighteenth and early nineteenth centuries, but a practical device had to wait upon an understanding of the workings of the electromagnet. When this became available in the mid-1830s, workable systems were developed in Germany, England, and the United States. In the United States, the unlikely inventor was Samuel F. B. Morse, an accomplished and well-known artist who in spite of his talent was finding it difficult to win financially rewarding commissions. Ambitious for both fame and

Thomas A. Edison (1847-1931) listening to his phonograph

14.
Edison Phonograph
About 1878
Catalog No. 320551

Samuel F. B. Morse (1791-1872) with telegraph

15

16

15.
Telegraph
Samuel F. B. Morse
Poughkeepsie, New York
May 1, 1849
Patent No. 6420

16.
Morse Telegraph (Replica)
1844
Catalog No. 200002

17.
Printing Telegraph
Thomas Edison
Newark, New Jersey
July 1, 1873
Patent No. 140488

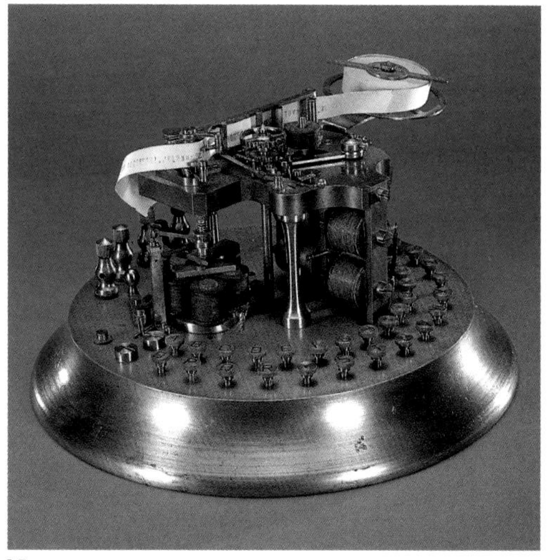

17

fortune, like so many others he was a dreamer. Unlike most others, he was extraordinarily successful.

Morse's first crude telegraph dates from 1835. The model shown here was constructed by his colleague Alfred Vail; it accompanied the patent claims granted in 1849. In essential respects, it is identical to his first practical telegraph receiver, inaugurated in May 1844 when Morse sent the message (in a code of dots and dashes), "What hath God wrought?" from Washington, D.C., to Baltimore, Maryland.

Figure 16 is a reproduction of the 1844 telegraph. Note in particular the large electromagnets. Pulses of electricity would cause them to pull on a lever which would then press dots and dashes into a paper tape. Pressing the key in Washington would send short and long pulses of electricity through a wire to the other end in Baltimore where they would activate the large, sensitive coils shown underneath the platform. These coils acted as a relay, closing a switch and producing pulses in a circuit that included the electromagnets of the receiver.

Telegraph lines quickly spread across the American continent. By 1861 lines reached from New York to San Francisco, and by 1866 across the Atlantic to England and Europe. There were new job opportunities, and this industry became the training ground for many future electricians. One of these was Edison, who spent his early years as a telegraph operator. His first important patent was for improvements on a printing telegraph used especially for receiving stock quotations. The money he made from this invention allowed him to establish his own laboratory at Menlo Park, New Jersey, where he made many of his most famous inventions. This model illustrates the principles of Edison's patent of 1871. It could operate both as a transmitter and as a receiver.

18

19

18.
Fire Alarm
Moses Farmer & William Channing
Salem, Massachusetts
May 19, 1857
Patent No. 17355

19.
Alarm Telegraph
Edwin Rogers
Boston, Massachusetts
June 14, 1870
Patent No. 104357

20.
Bell Telephone
About 1898
Catalog No. 181852

The telegraph also had an impact on life within cities. Special networks made it possible to organize police and fire services much more efficiently than ever before. Physician and promoter William Channing (1820-1890) teamed with the well-known electrician Moses Farmer (1820-1893) to produce America's first practical fire-alarm system. Call boxes, placed throughout a city, would automati-

Alexander G. Bell (1847-1922)

20

21

21.
Telephone
Alexander G. Bell
Salem, Massachusetts
March 7, 1876
Patent No. 174465

22.
Telephone
Thomas Edison
Menlo Park, New Jersey
September 23, 1878
Patent No. 208299

23.
Edison Telephone
About 1877
Catalog No. 314896

24.
Long Distance Telephone
Anthony C. White
Boston, Massachusetts
November 1, 1892
Patent No. 485311

cally send coded messages to the central office when activated. The code would identify the location of the box. The Farmer-Channing system was first installed in Boston in 1852, and in many other cities in succeeding years. Improvements in fire alarm systems continued to be made, as with this alarm ringer by E. Rogers.

The field of telegraphy attracted many would-be inventors. Alexander Graham Bell (1847-1922), who had recently migrated with his parents from Scotland to Canada, was teaching speech to deaf pupils in Boston in the early 1870s. Like Morse, he knew little of electricity; but also like Morse, he was ambitious for the rewards that a successful invention would bring. He experimented with a form of "multiple telegraphy" in which several messages could be sent over a single wire at the same time. In the course of that work he discovered the principle of the telephone. There were others who had similar ideas during this period, but Bell alone saw the practical value of the invention. His first successful speech transmission was in March 1876, when he spoke to his associate between two rooms in a house, saying "Mr. Watson, come here, I want to see you." Within another year the telephone was being installed commercially.

These patent models nicely illustrate the basics of Bell's telephone. Speaking into one of them causes a piece of iron attached to the membrane to move near the wire coil, inducing a fluctuating cur-

22

23

24

25

26

Hoe "Lightning" 10-feeder press for newspapers, about 1860

25.
Machine for Making Paper Bags
Margaret E. Knight
Springfield, Massachusetts
October 28, 1879
Patent No. 220925

26.
Printing Press
Richard M. Hoe
New York, New York
July 24, 1847
Patent No. 5199

rent in the wire. This current is conveyed by additional wire to the receiver, where it causes the coil to produce a fluctuating magnetic field, moving the iron and the membrane, and so the air, in the ear of the listener.

Bell's telephone transmitter was powered directly by the voice and as a consequence it worked well only over short distances. Several people designed alternative transmitters in which the voice could change the resistance of some element in the circuit—usually by changing the pressure on pieces of carbon. Edison, working in his Menlo Park laboratory, was one of these inventors. In the end, he received patent credit for his work. In this patent model, fluctuating pressure on the membranes causes the pressure between the pieces of carbon to fluctuate, producing a very effective transmitter that is essentially the same as most used today.

Bell's receiver was also quite effective, and is in principle the same as most still in use. To get around the Bell patent claims, the ever-resourceful Edison invented the device shown here. (See Figure 23) Turning the handle caused a chalk drum to rotate while pressing against a metal stylus in the earpiece. The fluctuating electrical signals from the transmitter passed between the stylus and the drum, changing the friction and hence also the scraping sound between them. The result was a strong, if somewhat raspy, reproduction of the voice. The Edison receiver was introduced in England, where it was used for a short time before the Bell company and Edison company merged.

An important improvement on Edison's transmitter was achieved by an inventor named White in 1892. Many particles of carbon were packed together to increase the amount of amplification and the tone.

PRINTING

Between 1639, when printing was introduced from Europe, and 1800, the American printer's tools changed little. With the beginning of the nineteenth century, a growing market of readers encouraged printers to look for faster ways to print, and inventors to provide the new equipment. Iron replaced wood, levers and rotary motion replaced press-screws, and reciprocal motion engines replaced human power. Similar changes came about in trades allied to printing: by mid-century paper-making, ink-making, type-founding, and type composition had been mechanized, and photography and electrotyping had been co-opted into picture printing. As the new equipment entered the trade shops, the trades themselves changed, particularly that of the printer. Instead of one printer handling all kinds of work, there were now printers who specialized in newspapers, books, or job work. And soon there was also specialized equipment for each.

At first the new machines and practices were copied from European models, but by 1850 American presses were competing in the world market. The patent models in this exhibition include some of the most successful of these early American printing inventions.

The wooden press of the eighteenth century, which could print at a maximum rate of 240 sheets per hour, printed by squeezing paper and type between two flat surfaces. This press was succeeded in the 1820s by machines on which type on a flat bed was rolled under a cylinder. These cylinder presses could print 600 sheets an hour, a speed that was acceptable for book printers. But printers of newspapers serving perhaps one hundred thousand subscribers needed much faster machines. Cylinder presses had to pause in their cycle after each impression, while the flat type bed returned to its starting position. It was obvious to many that the next mechanical advance would avoid this stop-and-start motion by wrapping both type and paper around cylinders, which could then turn continuously.

The leading American press manufacturer, R.M. Hoe, found himself in competition with an English company, Applegath and Cowper, to build such a machine. Both had to find a way to attach rectangular type to the surface of the cylinder. The Applegath press, which was unveiled in 1848, solved the problem by using a many-sided polygon instead of a true cylinder, and attaching the columns of type to the flat surfaces of the polygon. Hoe's press, introduced the same year, used wedge-shaped strips of metal between the columns of type to lock them around the cylinder. Hoe's press, at first plagued with reports of type falling off the cylinder and into the machinery below, was the more successful of the two.

Hoe's type-revolving press was made in

27.
Printing Press
William H. Golding
Chelsea, Massachusetts
December 2, 1873
Patent No. 145101

28.
Printing Press
Thomas C. Kenworth &
Archibald McGregor
New York, New York
May 7, 1878
Patent No. 203465

29.
Hand Printing Press
James N. Phelps
New York, New York
November 2, 1858
Patent No. 21980

30.
Type Casting Machine
David Bruce, Jr.
Williamsburg, New York
November 6, 1843
Patent No. 3324

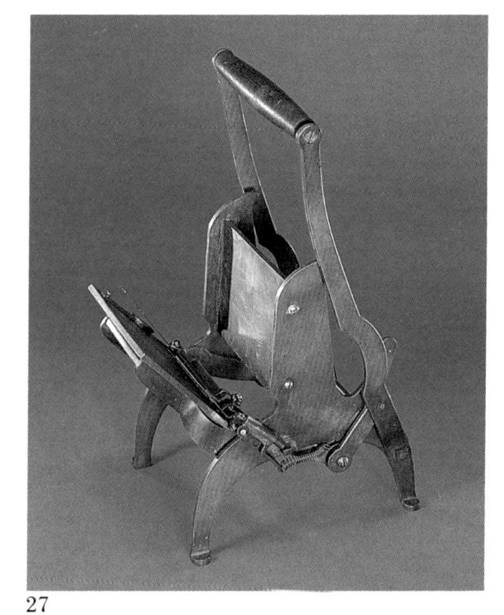

27

28

29

30

a series of sizes that had two to ten paper-feeding stations, each feeder contributing 2,000 sheets of paper an hour. The enormous ten-feeder press, known as the "Lightning," printed 20,000 sheets an hour. It was said to be as high as a three-story building and half a city block in length; it weighed some 30 tons and cost around $50,000.

Speed was not the only consideration for printers. "Jobbers" were printers who undertook such neighborhood work as tickets, invitations, posters, and billheads in runs ranging from ten or twenty to several hundred. The presses that were developed specifically for job printers were much slower in operation than newspaper presses, but they were comparatively quick to set up for each job. Powered by hand levers, cranks, or foot pedals, these presses were mechanically simple and easy to repair. They were ideal for a printer whose workday might include printing five different jobs in runs of one hundred each. William Golding of Boston patented and manufactured several of these jobbing presses between 1870 and 1900. This patent model was the basis for his "Pearl" line of presses.

From the middle of the century, two other kinds of printer entered the scene: the amateur and the businessman. Merchants, lawyers, pharmacists, and others who ran small businesses might want to print their own letterheads as needed, or to emboss documents with their office seal. The presses developed for the office were usually small enough to fit on a desk, and often handsomely decorated. They were neither powerful nor efficient, but they served the needs of their owners. The same presses were advertised for amateur printers, or as educational toys for boys. The Phelps press, which its inventor called the "Economist," would hold enough type for a name and address, and included an automatic inking system.

By the traditional method of type casting, practiced from the fifteenth century to the nineteenth, molten lead alloy was poured into a hand-held mold containing a copper matrix bearing the form of a letter. Making type this way was laborious, and too slow to supply the busy printers or the new composing machines of the nineteenth century. David Bruce of New York made the first successful attempt to mechanize the type-casting process. His machine, operated by a crank, imitated the motions of the type-caster's hands: molten metal was squirted into a mold, and a moment later the mold pivoted away from the metal pot and opened, dropping a newly cast piece of type. Bruce's invention is at the heart of some type-casting machines, known as "pivotal machines," used today,

31

32

31.
Camera
William Southworth
New Castle, Maine
June 17, 1862
Patent No. 35635

32.
Camera
John Stock
New York, New York
July 5, 1859
Patent No. 24671

PHOTOGRAPHY

During the nineteenth century, inventors in the United States made photography—a scientific and artistic novelty imported from France with the daguerreotype—into a wholly American industry. Very little of what the Americans invented was "new" in any radical or startling way; most of the photographic patent claims were improvements on existing technology. But the nature of these improvements, and the rapid pace at which photographic technology grew in the last half of the century, reflect the entrepreneurial spirit of nineteenth-century American inventors.

Could photography be made less expensive? More profitable? More convenient? There were many inventors out there ready to claim it could.

Who were these inventors? Most frequently they were manufacturers or designers working in other fields who saw in the "daguerrean arts" a new business opportunity (the Stock brothers of New

York, John and Jacob, claimed patents for barber chairs, among other things, before they turned their energies to camera equipment). In many cases these inventors saw a growing market for the new science, and went about getting a piece of it for themselves. Many worked for manufacturing firms and the patent records credit both the inventor and his employer.

Certainly the photographic inventions of the nineteenth century reflected advances, some quite significant, in mechanics, optics and chemistry—the basics of photography—but these advances were driven by economics. Many photographic inventions answered the photographer's needs: adjustable stands, easier-to-handle equipment, and mechanical improvements designed to produce a better picture. Some inventions allowed the photographer to make multiple images and others made it possible to reproduce images cheaply.

John Stock's 1859 patent specification claimed an improvement both in technology and convenience. The fitting for the lens tube on this basic daguerreotype camera body allows the lens to tilt in several directions without moving the rest of the camera. The plate frame attached to the back of the camera, in which the chemical-coated plate that was to hold the image was placed, could be moved up or down and from side to side. In this way, the plane or angle of the plate could be made to correspond to the plane of the subject being photographed, eliminating a certain distortion in the final picture.

The 1862 specification for William Southworth's "Multiplying Camera" claims an improvement in making multiple copies of a single image, through use of a sliding plate that would move the lenses. In this way the image could be projected onto areas of the plate chosen by the photographer.

As photographers began to travel more in the expanding United States, field kits or portable darkrooms were invented. The development of tintypes made photography cheaper—portraits cost only twenty-

33

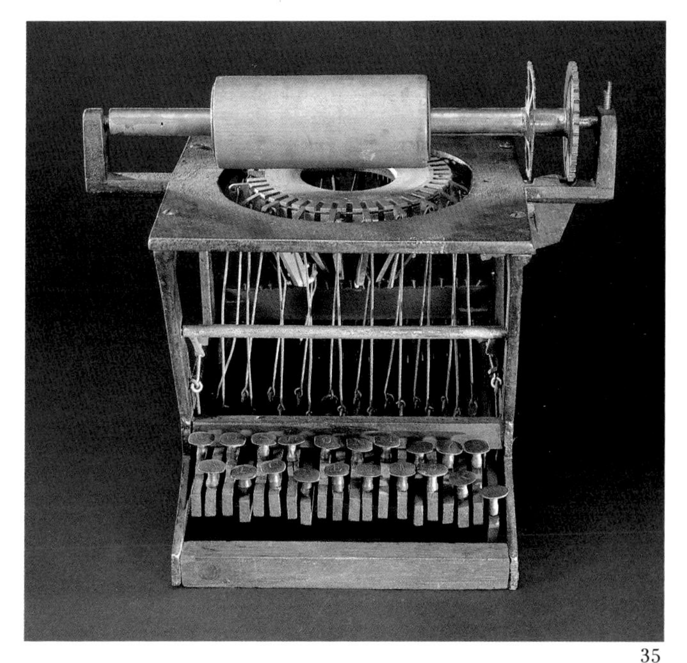

35

34

33.
Remington No. 1 Typewriter
About 1878
Catalog No. 181132

34.
Typewriter
Hans R.M.J. Hansen
Copenhagen, Denmark
April 23, 1872
Patent No. 125952

35.
Typewriter
C. Latham Sholes &
Matthias Schwalbach
Milwaukee, Wisconsin
September 19, 1876
Patent No. 182511

five cents from a roving tintypist—and so the traveling photographer became a common sight in the American West and at fairs and carnivals everywhere. At the same time, the sentimentality, home-centeredness, and penchant for decoration which distinguished the Victorian-era consumer in America created a ready retail market for photography, whether as portable, personal art (a miniature of a loved one) or a necessary item in the parlor (a stereoscope or photography album). Many of the patent models reflect these trends: elaborately decorated cases, albums, and frames, or such novelties as processes to render photographs three-dimensional.

In the National Museum of American History's collection there are nearly four hundred patent models reflecting the first forty years of photography in the United States; some are curiosities in today's world of high-tech instant photography, and others help us trace the development of photography as a science. But their value may lie mostly in what they say about the enterprising spirit of the nineteenth-century American scene.

TYPEWRITING

The modern typewriter originated in Milwaukee, Wisconsin, in a series of inventions beginning in 1867. Printer Christopher Latham Sholes and his collaborators received two patents for a writing machine in 1868. They persuaded Pennsylvania oilman James Densmore to help them finance the undertaking and organize manufacturing. Their efforts led to an 1872 prototype called "the Type Writer."

Densmore convinced the firm of E. Remington of Ilion, New York, to redesign, produce, and market the machine. Manufacturing began in 1874, and in its first year the firm sold about four hundred machines for $125 each. At first there was only a tiny market, and a significant group of users had yet to be identified. Typewriter sales picked up in the early 1880s, however, and competing manufacturers entered the field. Although many machines of radically different design appeared briefly on the market, by the turn of the twentieth century most machines, even those manufactured by other firms, looked and functioned almost like the Remington.

The National Museum of American History has an unparalleled collection of patent models of early writing machines. The selection offered here includes one of Christopher Sholes's models, which accompanied a patent application he and Milwaukee clockmaker Mathias Schwalbach filed for improvements on their Type Writer in 1876. The Remington typewriter was not the first writing machine to enjoy commercial success. The first to be sold in any quantity was the invention of a Danish pastor, Hans Johan Rasmus Malling Hansen. Hansen's Writing Ball was manufactured beginning in 1870, and the example in the Museum's collection dates from about that time. Hansen received a U.S. patent for the device in 1872; he submitted this model with his application.

36.
Hansen Typewriter
About 1872
Catalog No. 181005

Mechanical Solutions

Agriculture

When President Thomas Jefferson took office in 1801, the country was just twelve years old and its territory stopped roughly at the Mississippi River. A hundred years later, after various purchases and conquests by war, the boundaries had stretched to the Pacific Ocean and beyond, to Hawaii and to Alaska. Millions of farmers settled and then tilled this vast expanse of land.

In the early years of the nineteenth century, many farmers worked small farms and used draft animals and hand implements. In the southern states, rice, cotton, and tobacco plantations depended upon slave labor, while yeomen tended smaller farms and used family labor. The cotton gin, by economically and quickly separating the cotton lint from the seed, revolutionized southern agriculture, and the cotton culture—and slavery—spread rapidly across the South. Because the gin was such a practical and easily understood (and copied) invention, Eli Whitney never managed to protect his patent. Gin manufacturers proliferated as cotton cultivation expanded. But field workers on cotton farms, as well as on tobacco and rice farms, continued to rely upon draft animals and hand implements. A. W. Washburn's 1866 cotton-seed planter suggests that the inventor, a resident of Yazoo City, Mississippi, hoped to save labor by using the machine to prepare the seed bed and to drop the seed instead of doing it by hand. His initiative may have come from the fact that slavery had just been abolished during the Civil War.

Farmers throughout the country continually tinkered with machines that might save them labor. When Cyrus McCormick and Obed Hussey came up with the principles of the reaper, they set the stage for a series of inventions that over time led to a grain binder and evolved into the self-propelled combine. Most inventions, however, were not earthshaking; most pertained to small improvements in this transition from draft animals and hand labor to self-propelled machinery.

Praul's traction engine suggests a modern style of agriculture that relies upon technology rather than hand labor or draft animals. Windmills, used primarily in the Midwest and on the Great Plains, turned wind into energy that pumped water, again saving hand labor. Although dog-powered or even draft-animal treadmills could drive light farm machinery, steam- and later kerosene- and gasoline-powered machinery greatly expanded the available power. Steam engines at first drove rice mills, sugar refineries, and cotton gins. With the advent of steam traction engines, threshing and heavy plowing could be mechanized, but these heavy and cumber-

Men, possibly the owners and employees of the Beardsley Hill Company, stand behind an Auburn self-raking reaper at the factory, about 1860-1870.

37

38

37.
Animal Trap
George F. Lampkin
Georgetown, Kentucky
July 9, 1872
Patent No. 128802

38.
Animal Trap
A. A. Fradenburg
Nevada City, California
May 22, 1866
Patent No. 54885

39.
Cotton Seed Planter
A. W. Washburn
Yazoo City, Mississippi
March 26, 1866
Patent No. 14529

40.
Traction Engine
John E. Praul
Washington, D.C.
November 4, 1879
Patent No. 221354

39

40

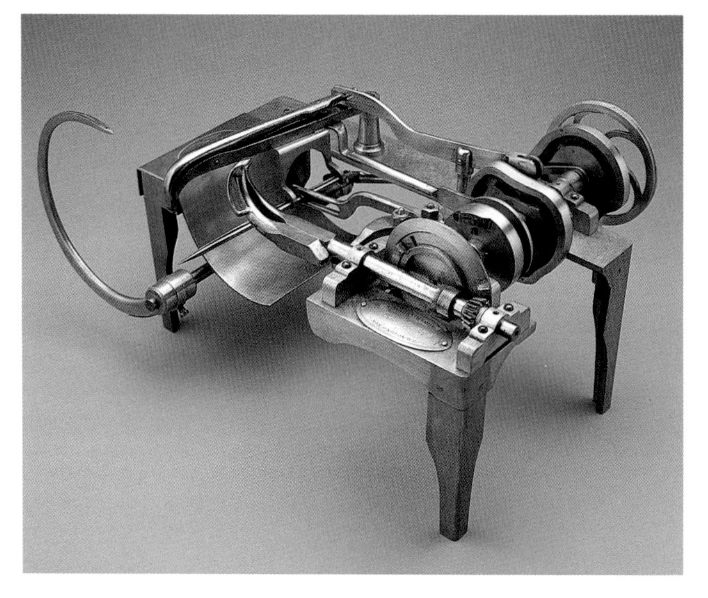

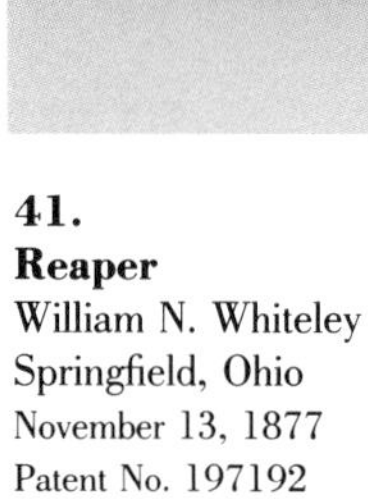

41.
Reaper
William N. Whiteley
Springfield, Ohio
November 13, 1877
Patent No. 197192

42.
Grain Binder
Harvey R. Ingledue
Gray, Iowa
February 3, 1885
Patent No. 311492

A reaper cutting grain around 1890

43.
Plow
Augustus G. Christman
Sheridan, Pennsylvania
January 20, 1880
Patent No. 223666

44.
Planter
Charles G. Everet
Bellefontaine, Ohio
July 13, 1880
Patent No. 229985

some machines were not well adapted to field work. Not until the twentieth century did small and agile farm tractors ease most of the plowing chores.

A. G. Christman's all-purpose plow of 1880 showed his desire to conquer all of the improbables of field work in his native Pennsylvania. His plow was suitable for all soils and conditions, and the shares and moldboard could easily be changed to accommodate prevailing conditions.

In 1880, C. G. Everet of Ohio patented

a corn planter that would allow the farmer to plant his seed by the check-row method, uniformly spacing the stalks a certain distance apart. Another Ohioan, G. W. Brown, invented a corn planter in 1883 that could avoid field obstacles such as rocks and trash by allowing the farmer, without leaving the machine, to trigger a spring-loaded mechanism that raised the runners.

P. F. Wells, a Michigan resident, patented an improved metal frame in 1882 that boasted better strength and convenience for the operator of a cultivator. New Yorker G. B. Davidson's 1892 cultivator allowed a greater range of tooth adjustment that could change in both the depth and angle of inclination. Such small improvements in existing machinery further eased the work of farmers.

Half a century after the invention of the reaper, Illinois resident J. H. Jones's improved mowing machine took advantage of a chain, bell-crank, eyebolts, and pulleys to trigger a device to raise or lower the cutter bar, in order to discharge grass or evade rocks or other obstacles. This mechanical innovation continued on mowers until hydraulics further eased operation.

Processing also necessitated invention, and Pennsylvanian J. H. Heinz's 1879 vegetable sorter, utilizing the principle of graduated drums, could easily separate different sizes. As vegetables moved through the first drum, the smaller size fell out and the process was repeated until the sorting was accomplished. Iowa resident J. A. Wade's 1879 flax-seed separator boasted "improvements in machines for separating flaxseed from chaff and other foreign substances; and the invention con-

45

46

47

45.
Corn Planter
George W. Brown
Galesburg, Illinois
May 29, 1883
Patent No. 278497

46.
Cultivator
Philip F. Wells
Milford, Michigan
January 24, 1882
Patent No. 252637

47.
Mowing Machine
James Herva Jones
Rockford, Illinois
June 23, 1891
Patent No. 454741

sisted in the construction and novel arrangement of parts."

There were thousands of similar inventions. These models symbolize an accumulation of mechanical steps that eased farm work. At the same time, such technology encouraged the growth of larger farms to utilize the additional power. Farm work in the nineteenth century relied in most cases upon a structure that alloted labor and shared costs within a community. Highly mechanized work such as harvesting, shocking, and then threshing small grains actually created groups of farmers, usually called threshing rings, that preserved a sense of neighborhood and community. Technological change hardly affected such southern crops as cotton, tobacco, and sugar—except at the processing stage—nor would major changes transform southern agriculture until the middle of the twentieth century.

It is important to understand that most inventions in the nineteenth-century United States originated from practical tinkerers. In that age, large laboratories and research budgets were not required to contribute the incremental improvements that eased farm labor. Ultimately, invention moved inside large corporations, and farmers increasingly relied upon ideas that flowed from corporate science and technology. Such inventions, by saving labor and permitting larger farms, drove many farm families from the land. In our own time this paradox has led historians and others to question the wisdom of labor-displacing technologies.

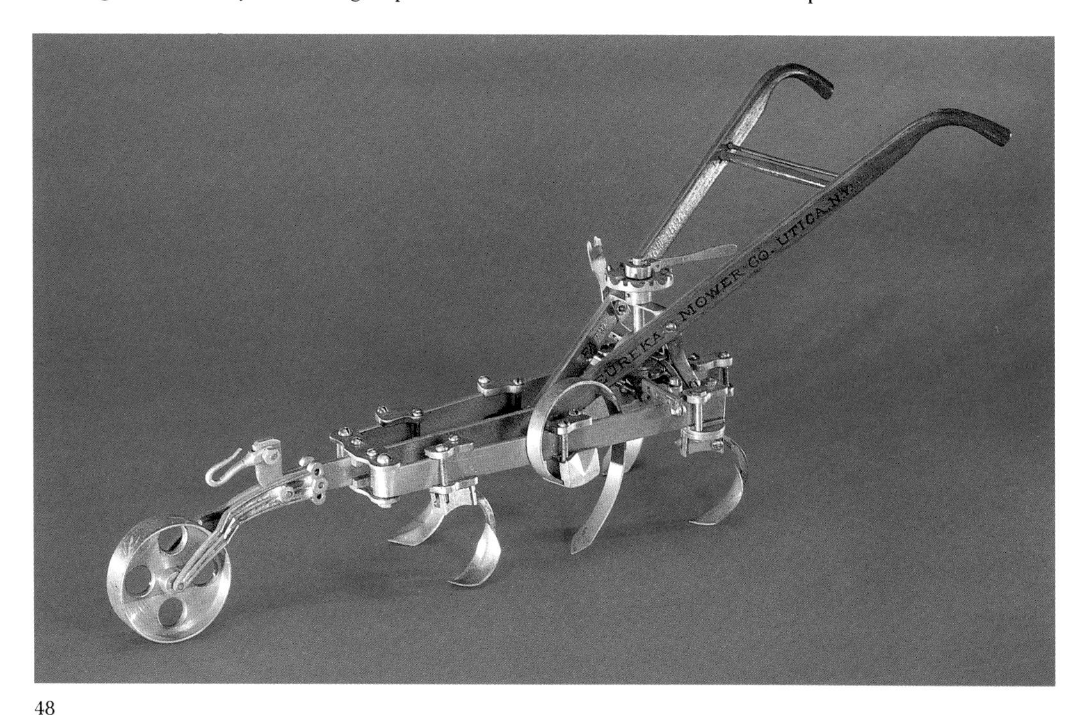

48

49

50

48.
Cultivator
George B. Davison
Utica, New York
January 26, 1892
Patent No. 467660

49.
Vegetable Sorter
John H. Heinz
Sharpsburg, Pennsylvania
February 4, 1879
Patent No. 212000

50.
Flax Seed Separator
Jeremiah A. Wade
Cherokee, Iowa
September 30, 1879
Patent No. 220211

TEXTILES

The earliest mechanized devices for spinning and weaving in the United States were imported from England at the end of the eighteenth century. Throughout the 1800s, patented improvements to looms and spinning machinery revolutionized the structure of America's system of textile manufacture. These improvements in turn allowed a greater use of textiles by people of modest means. The production of household goods such as sheeting, drapery and upholstery fabrics, and floor carpeting became increasingly mechanized. By the end of the nineteenth century, household fabrics were cheaper than they ever had been, and were available in a greater variety of prints, weaves, and textures than ever before.

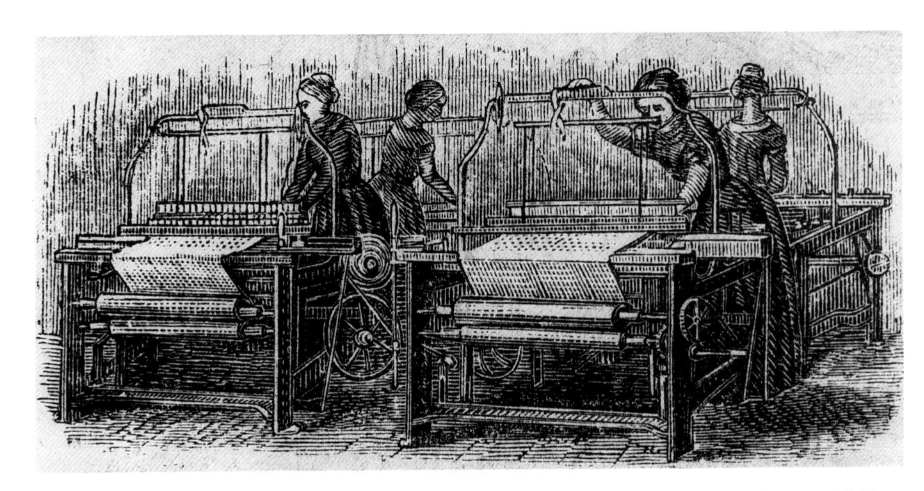

Iron frame looms shown in an advertisement by Sumner Pratt of Worcester, Massachusetts, 1860

The first patented invention to bring about a radical change in America's textile industry was Eli Whitney's cotton gin of 1794. By mechanizing the initial stage of cotton production, this device ultimately made the United States an exporter of raw cotton instead of an importer.

While the cotton gin was a new way of solving a processing problem, most inventions pertained to saving labor at the production stage through incremental improvements in speed or efficiency. The power loom and mechanized spinning machinery, which reached the United States soon after their introduction in England, provided a basis for numerous subsequent improvements. England officially prohibited the export of machinery until 1845, and one effect of this was to force Americans to devise their own improvements. Most of these dealt with specific American needs. The ring spindle, used in Britain rather than the Mule, was adopted for spinning cotton because it was more productive per operator and could be operated by the people who made up our work force.

The chronic scarcity of labor in the nineteenth century was definitely a factor in encouraging American invention. One English observer noted that American workmen "hailed with satisfaction all mechanical improvements, the importance of

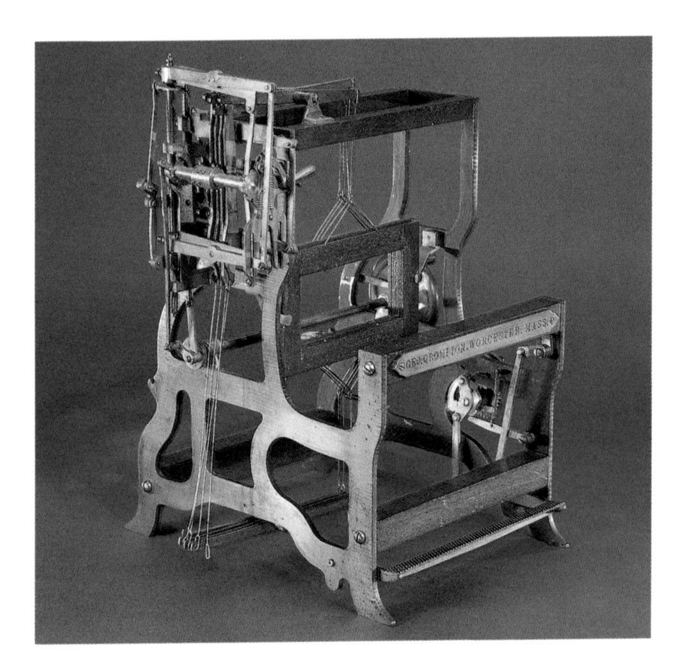

51.
Loom
George Crompton
Worcester, Massachusetts
September 7, 1869
Patent No. 94571

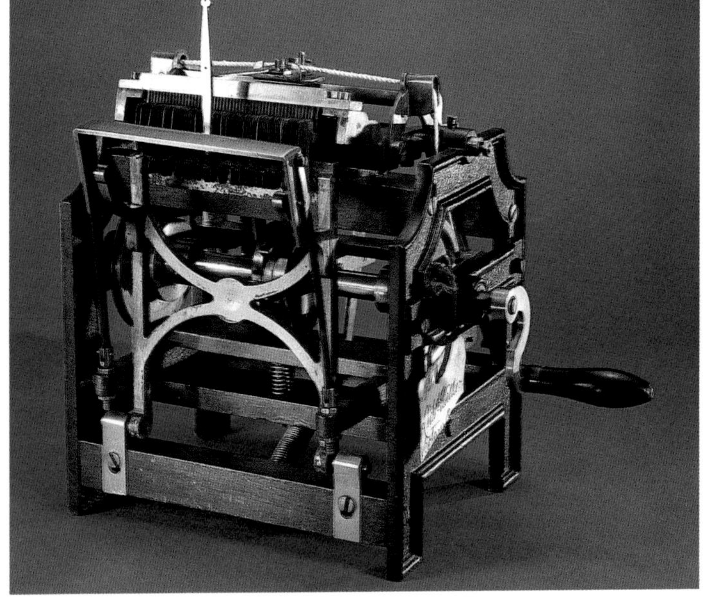

52.
Knitting Machine
Richard Walker
Portsmouth, New Hampshire
December 5, 1839
Patent No. 1421

53.
Eli Whitney Cotton Gin
Courtroom model
About 1800
Catalog No. T.8756

54.
Carpet Loom
Erastus B. Bigelow
Boston, Massachusetts
May 30, 1876
Patent No. 177920

55

55.
Carding Machine
Hiram Houghton
Somers, Connecticut
April 21, 1857
Patent No. 17094

56.
Design for Carpet
Henry G. Thompson
New York, New York
August 7, 1860
Design Patent No. 1306

which, as releasing them from drudgery of unskilled labor, they were enabled by education to understand and appreciate." Immigration increased later in the century, augmenting the supply of unskilled labor, but skilled labor remained scarce. Machines that required relatively little training to operate and monitor were essential to the continued expansion of the textile industry.

Finally, the types of goods produced in the United States also influenced inventors of textile machinery. Because Americans continued to import silk and linen goods, less attention was devoted to improvements in those industries. Coarse, plain weaves of cotton were the most widely produced cloths. The Englishman James Northrup, who invented the first fully automatic power loom for plain weaving, was therefore welcomed to the United States in the 1890s, along with his machine.

As noted earlier, one of the greatest accomplishments of the American textile industry in the nineteenth century was to make available cheaper and more varied household fabrics. Only the very wealthy, for example, could afford carpeting. But Erastus Bigelow and other inventors focused on reducing the amount of skilled labor entailed in carpet-weaving and increasing the productivity of each weaver by applying power to Jacquard-equipped looms for weaving ingrain and Brussels carpeting. Altogether, forty of his fifty patents were improvements to looms. Bigelow and his fellow inventors contributed to the vitality of an American textile industry that continues to foster innovations on the threshold of the twenty-first century.

SEWING MACHINES

As early as the mid-eighteenth century, inventors in England, France, Scotland, Germany, and Austria sought to overcome the technical problems of mechanical sewing. Often they tried simply to mimic the motions used in hand sewing, with no success. Some American inventors who looked for ways to mechanize sewing in the early nineteenth century also attempted to simulate the motions of sewing by hand, but other took a more innovative approach. Elias Howe, Jr., patented his first sewing machine in 1846. It incorporated some, but not all, of the elements of a successful machine. Other inventors continued the quest.

A combination of their patented improvements ultimately resulted in several practical sewing machines. The necessary features were an eye-pointed needle; continuous thread from spools; a horizontal table; a lock stitch; a shuttle or bobbin for a second thread; an overhanging arm; synchronous cloth feed and needle motion; a presser foot; and a capability for either curved or straight stitching. These individual improvements, along with the mechanisms to insure that each operation was carried out in proper sequence, comprised the fundamentals of a practical sewing machine.

In 1851 three sewing machine companies were formed on the basis of patent rights owned by the founding inventors

57.
Spinning Machine
Charles Danforth
Paterson, New Jersey
December 12, 1854
Patent No. 12055

58.
Sewing Machine
Job A. Davis
New York, New York
February 21, 1860
Patent No. 27208

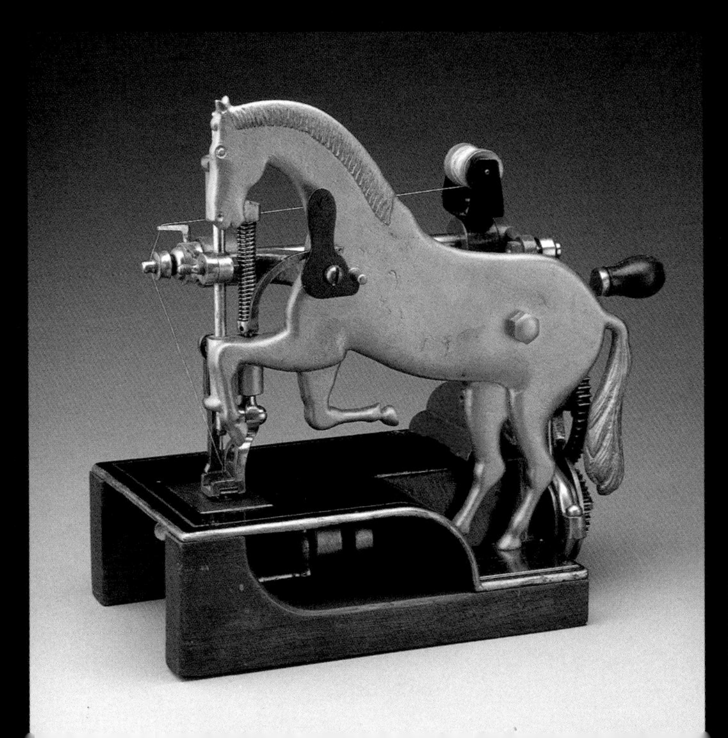

59.
Sewing Machine
James Perry
New York, New York
November 23, 1858
Patent No. 22148

This trade card depicts an eager family awaiting the delivery of their Wheeler and Wilson No. 8 sewing machine, about 1890.

OLD

Wheeler and Wilson advertisements contrast the "old" way of sewing by hand to the "new" way of sewing with the sewing machine, implying the subsequent improvement in life.

NEW

60.
Sewing Machine
David W. Clark
Bridgeport, Connecticut
July 5, 1858
Patent No. 19015

61.
Sewing Machine
George Hensel
New York, New York
July 12, 1859
Patent No. 24737

60

61

and their partners. These were the Wheeler & Wilson Co., the Grover and Baker Co., and I. M. Singer & Co. Each company produced a practical sewing machine and seemed to understand how to market a radically new product to the nineteenth-century consumer. In 1856, after years of lawsuits over patent rights, Elias Howe and these three companies formed the Sewing Machine Combination and joined their patents in the first patent pool in American industry. This freed the companies from expensive and time-consuming litigation and enabled them to concentrate on manufacturing and marketing their machines.

The sewing machines of the 1850s were often too cumbersome and expensive for many consumers, so their use was confined to the manufacturing trades. Wheeler & Wilson machines, which were light (6 lbs. as compared to Singer's 53 lbs.), introduced the sewing machine as a home appliance. Grover received a patent in 1856 for the first portable case for a sewing machine. Machines for domestic use were often decorated with hand-painted flowers, gold leaf, and mother-of-pearl. They were available as table-top or treadle models, or encased in wooden cabinets that disguised their function and presented the machine as a piece of furniture.

As inventors overcame technical problems of machine sewing, some sought to further enhance the appearance of the machines. This presentation of the machine as an ornament was more attractive to Victorian sensibilities. Some inventors envisioned the machines as art, producing ideas for machines patterned after animals, fictional characters, or natural flora. Few of these machines were actually produced.

Marketing innovations helped ensure the success of the sewing machine as a home appliance. The Singer Company pioneered the concepts of installment sales, showrooms staffed with experienced operators and repair mechanics, and advertising geared specifically to the domestic market. Other companies were not far behind.

The sewing machine also had a significant impact on the ready-made clothing industry. Clothing could be produced at a much faster rate and sold for a lower price. In 1861, a man's dress shirt sewed by hand required 14 hours and 26 minutes to complete. On a Wheeler and Wilson sewing machine, the identical shirt took only 1 hour and 16 minutes to complete. Clothing styles became more elaborate and varied.

By the 1870s, with demand for their machines steadily increasing, some manu-

62

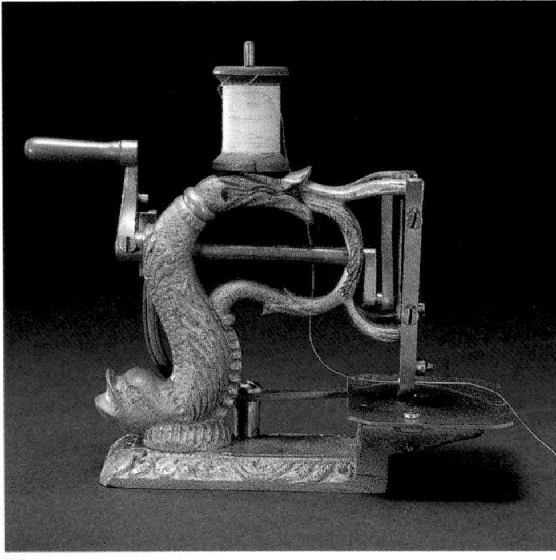
63

64

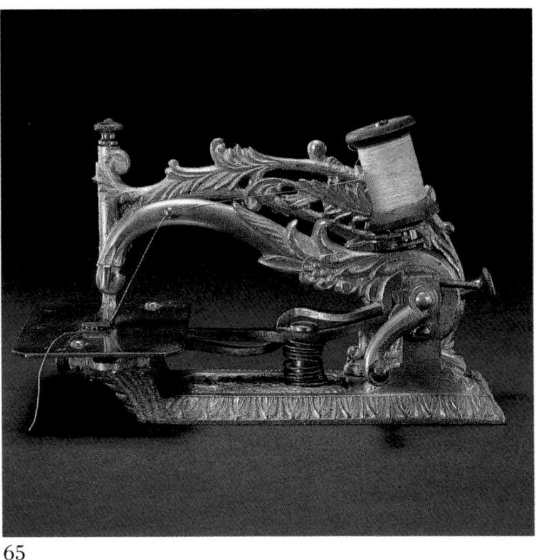
65

62.
Sewing Machine
Allen B. Wilson
Watertown, Connecticut
June 15, 1852
Patent No. 9041

63.
Sewing Machine
David W. Clark
Bridgeport, Connecticut
January 19, 1858
Patent No. 19129

64.
Sewing Machine
William O. Grover
Boston, Massachusetts
May 27, 1856
Patent No. 14956

65.
Sewing Machine
David W. Clark
Bridgeport, Connecticut
August 31, 1858
Patent No. 21322

facturers began to realize the importance of interchangeable parts. Soon factories were using many automatic and semi-automatic machine tools and employing an extensive gauging system for quality control. The resulting sewing machines were faster and quieter; they also stitched more efficiently, for they incorporated cranks, levers, and cams rather than bevel gears. By 1882, Knight's American Mechanical Dictionary illustrated 68 different sewing-machine stitches in use.

66.
Sewing Machine
Isaac M. Singer
New York, New York
August 12, 1851
Patent No. 8294

Isaac M. Singer
(1811-1875)

ISAAC SINGER

For his first patent model, Isaac Singer submitted a commercial sewing machine. He was granted Patent No. 8,294 on August 12, 1851. These commercial machines were built by hand in Orson C. Phelps's machine shop in Boston, Massachusetts. The head, base, cams, and gear wheels of the machine were made of cast iron; to fit together, these pieces had to be filed and ground by hand. The machine made a lockstitch by using a straight, eye-pointed needle and a reciprocating shuttle. The specific patent claims allowed were for: 1) the additional forward motion of the shuttle to tighten the stitch; 2) the use of a friction pad to control the tension of the thread from the spool; and 3) placing the thread spool on an adjustable arm to permit thread to be used as needed.

Always the showman, Singer relished exhibiting his invention at social gatherings and was good at convincing the women present that the sewing machine was a tool they could learn to use. The machine was transported in its packing crate, which served as a stand; it contained a wooden treadle that allowed the seamstress to power the machine with her foot, leaving both hands free to guide the cloth. This early, heavy-duty Singer machine was designed for use in the manufacturing trades rather than in the home.

Singer's success was due to several factors. First was the introduction of the hire-purchase plan (installment buying) by Edward Clark, Singer's partner and lawyer. A family could buy a sewing machine by putting a small amount down and paying the balance in monthly installments. This option was especially important in a time when few families could afford to buy a sewing machine outright. Second, with the introduction of the "Family" model, Singer started to advertise the sewing machine specifically for the home and sales proved the value of this aggressive approach. Third, Clark instituted a nationwide franchise system of stores staffed with sales agents, young women to demonstrate the machines, and mechanics to service and repair them.

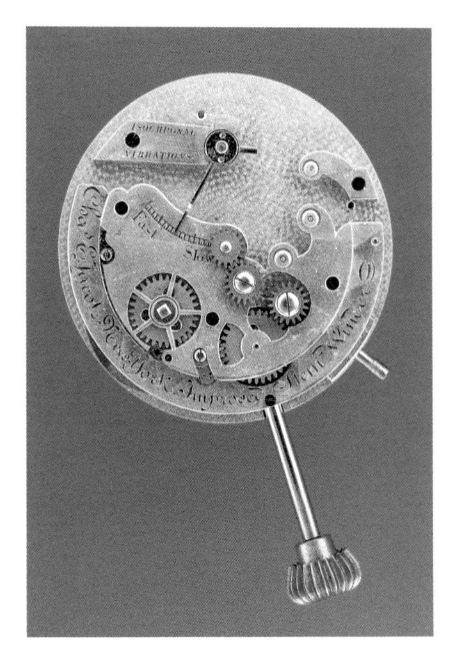

67.
Pocket Watch
Charles E. Jacot
Chaux-de-Fonds, Switzerland
September 27, 1864
Patent No. 44493

68.
Pocket Watch
Daniel Azro A. Buck
Waterbury, Connecticut
December 16, 1879
Patent No. 222658

WATCHMAKING

The watch was invented somewhere in Western Europe in the fifteenth century when an anonymous clockmaker substituted a wound spring for the driving weights in his timepiece. That spring made the miniaturization of timekeepers possible. From that moment until the middle of the nineteenth century, craft workers made watches by hand with the aid of specialized tools. Precise division of labor characterized the process, in which an individual worker, most often in a home workshop, performed a particular stage of production or finished a single part of the watch. British and Swiss firms dominated the world market.

Watchmaking underwent a revolution in the United States in the mid-nineteenth century. In 1849 Aaron Dennison at the American Watch Company in Waltham, Massachusetts, first attempted to produce watches almost completely by machine. Until then, there had been virtually no watch industry in the nation. Mechanization of watchmaking began slowly because watches were difficult to make by machine. There were many tiny parts, and the tolerances for error were even smaller. But after a decade of experimentation, Waltham had great success with machine-made watches. An enormous market of eager consumers awaited the arrival of these mass-produced American watches. Manufacturing in quantity lowered prices, and the market expanded in unprecedented ways. People who had never owned a watch before were able to afford one for the first time.

Waltham's success inspired competition. Howard, Elgin, Illinois, Hampden, Waterbury, Hamilton, and eventually some seventy other American firms emulated Waltham's methods and products. Watchmaking became one of the proving grounds of the American Industrial Revolution, in which goods were produced according to what the English called the "American system of manufactures." By the end of the nineteenth century, the fully developed system featured specialized machinery and precision gauges used in making products with interchangeable parts, in vast quantities.

Making watches by machine called for new watch designs. The "full-plate" movement Aaron Dennison put into production set the industry standard for roughly its first twenty years. By the 1870s, as other firms began producing watches, mechanics flooded the industry with innovations. Documenting this inventive activity are these two patent models. The dial patent was issued to Daniel Azro Ashley Buck, a New England watchmaker, during his tenure as the mechanical genius behind the Waterbury Watch Company. Waterbury's products were the cheapest American watches then available—under four dollars. A significant change in watch design was the switch from key wind to stem-wind and set. The Swiss introduced stem-winding beginning in the 1840s, but the American industry did not adopt the change on a large scale until the 1870s. The patent for a stem-winding watch was granted to Charles E. Jacot, a Swiss watchmaker from La Chaux-de-Fonds.

69

70

71

69.
Gear-Cutting Machine
John A. Peer
San Francisco, California
July 21, 1874
Patent No. 153370

70.
Candy-Making Machine
Thomas and George Mills
Philadelphia, Pennsylvania
February 14, 1871
Patent No. 111765

71.
Pill-Coating Machine
William Cairnes
Jersey City, New Jersey
February 16, 1875
Patent No. 159899

OTHER MANUFACTURES

The mechanization of textile production lay at the very heart of the Industrial Revolution in America. Sewing by machine, initially part of an industrial process, became something quite distinct as a result of the entrepreneurial genius of Isaac Singer and others—a domestic process as well, the beginning of "the industrialization of the home," in Ruth Cowan's phrase. The sewing machine, along with the mass-produced watch, also became emblematic of the consumer-goods revolution, a revolution that has had a profound impact on American history and on world history. After mid-century, and particularly after the Civil War, manufacturers of precision-fit articles began to strive for a degree of uniformity that would permit assembly from components selected at random, thereby greatly speeding the production process. The concept of "interchangeable parts," brought to perfection early in the twentieth century by Henry Ford as a key element of his system of "mass production," was central to bringing the price of every sort of consumer durable goods within reach of ordinary citizens. And that represented a social revolution as profound as any political revolution.

Essential to the manufacture of interchangeable parts were precision gauges and precision machine tools. In the years just before the model requirement was rescinded by the Patent Office, models for metalworking machines—"machines to make machines"—were often submitted that were perfect working miniatures. A fine example is the model that accompanied John Peer's application for an improved gear-cutting machine, Patent No. 153,370 of July 21, 1874.

Yet all consumer-goods production is not all mass production, not even now, and in the nineteenth century, inventors continued to develop devices that are best seen as something midway between handicraft methods and mechanization. One realm in which such devices remained commonplace was the processing of food and pharmaceuticals; two good examples are the Mills candy-making machine, Patent No. 111,765 and the Cairns "Apparatus for Sugar-coating Confectionary, Pills, etc.," Patent No. 159,899. And, at still another level, inventors continued to come up with tools and devices that were entirely in keeping with handicrafting—implements for carpenters, blacksmiths, shoe repairmen, and hundreds of other kinds of artisans. Mechanization is said to have "taken command" in the nineteenth and twentieth centuries. And so it may have, to a very considerable extent. But one should also consider that some of the tools and devices seen here remain the stock in trade of artisans who have never been displaced by mechanization, and quite likely never will be.

72

73

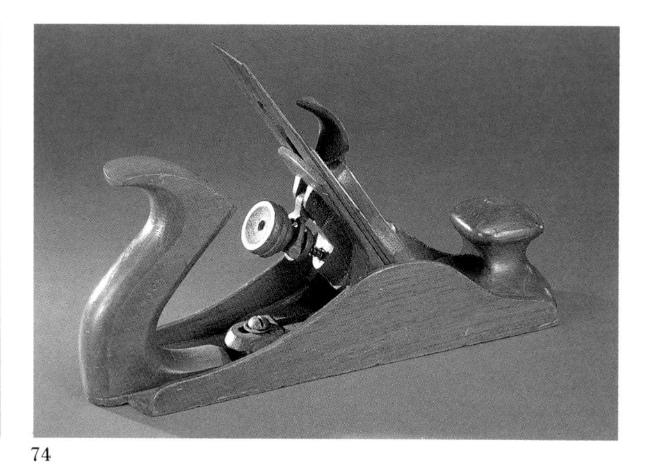

74

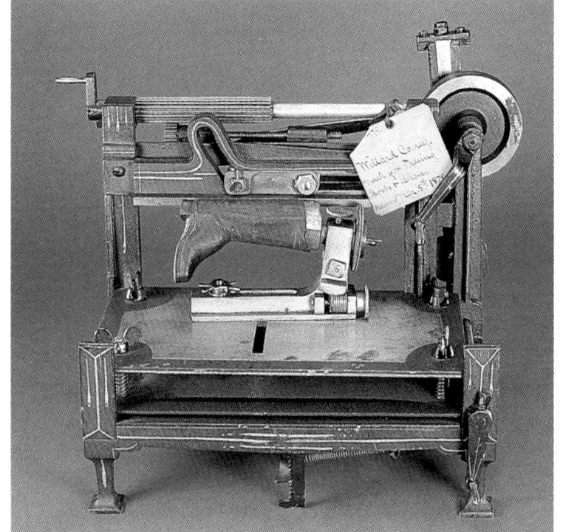

75

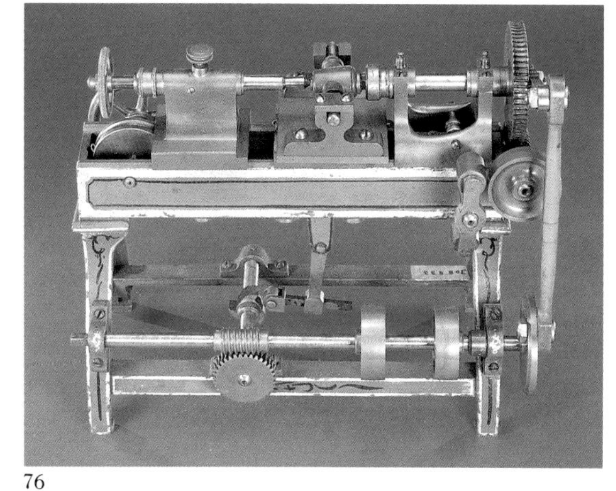

76

72.
Adding Machine
Jabez Burns
New York, New York
August 24, 1858
Patent No. 21243

73.
Adding Machine
Gustavus Linderoos
Point Arena, California
June 24, 1873
Patent No. 140146

74.
Wood Plane
Henry B. Price
New York, New York
June 17, 1879
Patent No. 216698

75.
Machine for Treeing Boots & Shoes
Willard Comey
Westborough, Massachusetts
March 5, 1878
Patent No. 200979

76.
Grinding Machine
Joseph L. Hayden
Cambridgeport, Massachusetts
January 13, 1880
Patent No. 223507

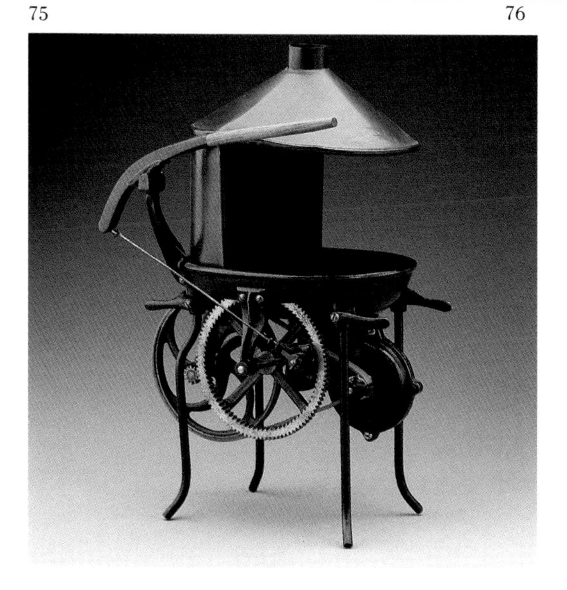

77.
Portable Forge
Charles Hammelmann
Buffalo, New York
January 10, 1882
Patent No. 252103

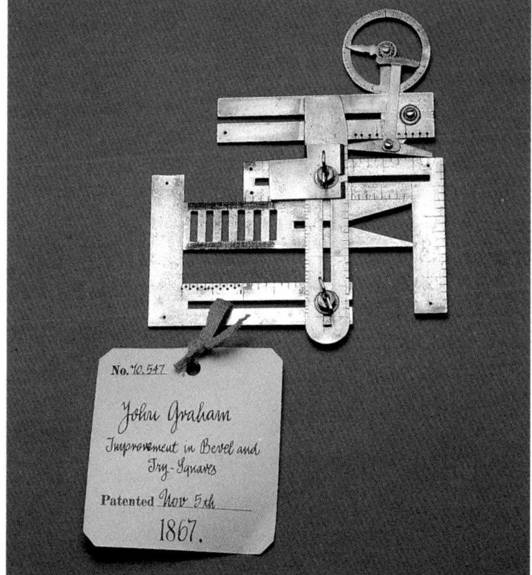

78.
Combination Tool
John Graham
Ludlow, Vermont
November 5, 1867
Patent No. 70547

At Home and at Work

★ Household Furniture

★ Lighting Devices

★ Electric Lights

★ Steam, Wind, Hot Air, and Internal Combustion Engines

★ Electricity

Masthead from journal promoting female invention, April 1890

Household Furniture

Patents for furniture reflect some of the concerns of nineteenth-century homemakers—their interest in comfort, for example, and their desire for more spaciousness in their living quarters. Occasionally the patents provide an unexpected glimpse of daily life or tell us something about contemporary customs. Whatever the case, most furniture patents pertain to minor improvements. After deciphering the specifications, examining the drawings, and studying the patent model, we usually learn that all the applicant is claiming is a better hook or brace or hinge.

Once letters patent were issued, however (and assuming an item of furniture was actually manufactured), the patentee might proudly mark the furniture "patented" and with equal satisfaction include that word in advertisements. A few women received patents for furniture. The question of just how many furniture patents actually formed the basis for commercial production remains to be determined. In the meantime, nineteenth-century patent models help us to envision the domestic surroundings—and appreciate the inventiveness—of ordinary citizens.

For families of moderate means with limited space, the patented pieces seen here filled a need and some provided a modicum of comfort. These were the people most likely to buy and use this furniture, and perhaps by improving it, to join the ranks of patentees themselves.

Between 1831 and 1905 the U.S. Patent Office issued more than three hundred

79

80

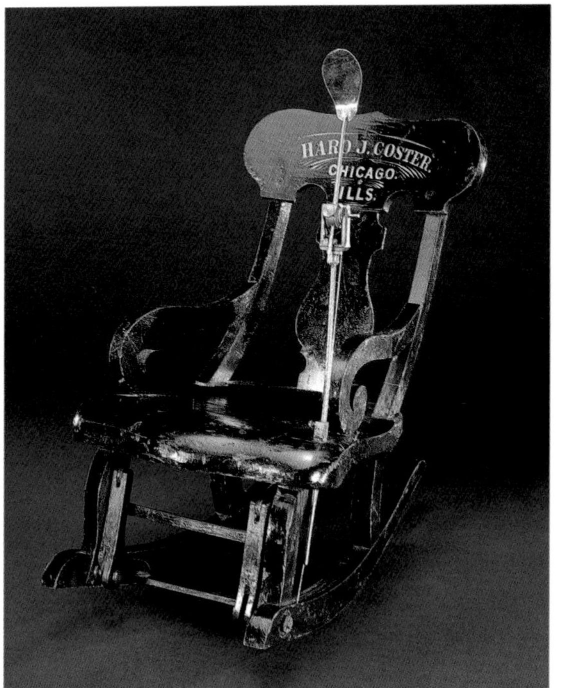

81

79.
Stove
Stephen Culver
Newark, New York
December 1, 1868
Patent No. 84537

80.
Oven
Alexander White
Geneseo, Illinois
February 8, 1870
Patent No. 99613

81.
Rocking Chair with Fan
Haro J. Coster
Chicago, Illinois
May 8, 1860
Patent No. 28159

patents for improvements in rocking chairs. For his "new and useful improvements in Rocking-Chairs," namely brackets and a fan, Haro J. Coster of Chicago, Illinois, received Patent No. 28,159 (May 8, 1860). To obtain the patent, which protected his invention for fourteen years, Coster submitted a model along with drawings and specifications of his invention. The model copies in miniature a type of chair popular at mid-century, made both with and without rockers and arms.

Attached to the front legs of the model are brackets that swivel up against the legs or down onto the floor, "provided with rollers . . . to prevent them from injuring the carpet or floor." The purpose of the brackets, Coster explained in the specifications, was "to hold the chair in the steady position" tilted back. This position permitted the occupant "to rest or sleep, without being obliged to hold the chair in such a position with his feet."

An inherent characteristic of a rocking chair is that it allows the occupant to lounge or tilt back. With Coster's invention this position could be easily maintained. That he felt it worthwhile to patent the brackets suggests that lounging was widely accepted, at least in an informal setting. In mentioning possible damage to the floor or carpet from his invention, Coster unwittingly gave us a peek at housekeeping practices and the extensive use of floor coverings in the mid-1800s.

The fan portion of Coster's invention was aimed at comfort, too. By circulating air, it made one feel cooler and possibly refreshed. The whirling motion might also keep irritating insects at bay, a worthwhile consideration at a time when windows and doors lacked screens. Neither the fan contraption nor the insect problem were new. A stationary chair with fan attachment worked by foot pedals was owned by no less a figure than President George Washington. Coster's claim, however, was to "provide the rocking chair with an automatical fan." Activated by motion, the fan continued to swing down and up "as long as the chair is rocking."

Beyond comfort, inventors were also attracted to space-saving devices. Between 1827 and 1873, more than 175 patents were issued for sofa, wardrobe, lounge, cabinet, and parlor beds. These multipurpose inventions could serve a number of functions in the space of one piece of furniture.

82

82.
Sofa and Bed Combined
Julius Werner
New York, New York
September 12, 1871
Patent No. 118994

83

83.
School Desk
Joseph Ingels
Milton, Indiana
May 10, 1870
Patent No. 102941

84A

84B

84A.
Folding Chair
Claudius & Nicholas Collignon
Closter, New Jersey
September 29, 1868
Patent No. 82494

84B.
Collignon Folding Chair
About 1870
Catalog No. 307707.4

85.
School Desk
Charles Perley
New York, New York
May 24, 1859
Patent No. 24151

86.
School desk
Sylvanus Cox & William Fanning
Richmond, Indiana
January 21, 1873
Patent No. 135089

87.
Playpen
Hiram J. Parker
Rochester, New York
November 2, 1875
Patent No. 169471

The model Julius Werner submitted when he applied for a patent demonstrates this New Yorker's ingenious method for converting a sofa into a bed. The back and seat flipped over to become the mattress. This was a refinement that would appeal to fastidious users; more often than not, bed and sofa shared the same upholstered area. The Werner arrangement also would lengthen the life of the piece by distributing the wear over both surfaces. After receiving Patent No. 118,944 (September 12, 1871), Werner had a circular printed up and distributed to illustrate his invention, explain how it worked, and seek agents to handle it. Because he was a furniture manufacturer, Werner's "improved patent parlor sofa bed" may have gone into production. If so, it could have been placed in the parlor, as Werner indicated by calling it a "parlor sofa bed." The term "parlor" gave status to the piece. The sofa bed was also to be found in the library and occasionally the hallway.

The many patents received by the Collignons—Adam, Claudius O., Nicholas, and Peter—pertained to the large variety of folding chairs they manufactured in their factory in Closter, New Jersey, and sold in their salesroom in New York City. A family business established in the late 1860s, Collignon Bros. featured the patents in their advertisements, leaflets, and catalogs for the remainder of the century. They also clearly marked their folding chairs. Some surviving examples have paper labels attached under the front seat rail reading "Collignon's Patent" and on the next line, the dates of pertinent patents, and below, "Closter, Bergen Co., N.J."

The models the Collignon brothers submitted with their patent applications usually duplicated in miniature their full-size chairs. Most of the patents pertained to folding actions. The chairs manufactured by Collignon Bros. and their many competitors—some of whose seat furniture also survives in quantity—folded flat for ease in shipping and storing and moving about from one room to another or even outdoors, wherever people needed to sit. Mass-produced and modestly priced, the folding chair appealed to almost every taste and pocketbook. The patent records reflect this popularity: Between 1855, when the first U.S. patent for a folding chair was issued, and the end of the century, more than 335 patents were granted.

Hiram J. Parker, an insurance agent,

88A

88B

88A.
Maytag Washing Machine
1915
Catalog No. 330502

88B.
Washing Machine
A. J. Stafford & S. Crossman
Essex, New York
January 9, 1866
Patent No. 51977

89.
Washing Machine
William Wheeler
Acton, Massachusetts
November 28, 1854
Patent No. 12012

90.
Refrigerator or Ice Box
Adam Heinz
Buffalo, New York
April 1, 1879
Patent No. 213751

89

90

91.
Clothespins

1. Dexter Pierce
   Sunapee, New Hampshire
   May 28, 1858
   Patent No. 20364

2. Jeremiah Greenwood
   Fitchburg, Massachusetts
   November 15, 1864
   Patent No. 45119

3. Henry W. Sargeant, Jr.
   Boston, Massachusetts
   April 11, 1865
   Patent No. 47223

4. T. L. Goble
   Orange, New York
   December 18, 1866
   Patent No. 60627

5. David M. Smith
   Springfield, Vermont
   April 9, 1867
   Patent No. 63759

6. A. L. Taylor
   Springfield, Vermont
   April 7, 1868
   Patent No. 76547

7. Henry Mellish
   Walpole, New Hampshire
   October 17, 1871
   Patent No. 119938

8. Vincent Urso & Benjamin
   Charles
   Evansville, Indiana
   August 12, 1873
   Patent No. 141740

9. Richard B. Perkins
   Hornellsville, New York
   February 20, 1883
   Patent No. 272762

salesman, and bookkeeper in Rochester, New York, was also a patentee. His invention for "improvements in inclosures or play-houses for children," was granted Patent No. 169,471 (November 2, 1875). Essentially it was a folding fence. "Closed into a fourfold compact block," it could be laid "aside for future use." Open, it became an enclosed rectangular space where, as Parker explained in the specifications, "in the absence of an attendant, small children may be placed in security to themselves and without constant attention from others." Mundane as the idea may seem to us, it may well have been novel at the time. Furniture of this type specifically for children and their care was a recent development. The "inclosure" most likely signals changing attitudes about children and their training. The space-saving feature of Parker's "playhouse for children" also was an important consideration.

Lighting Devices

The nineteenth century was a revolutionary period in lighting technology. For centuries there had been no change in the devices used to produce artificial light. The first truly scientific development in lighting was introduced by the Swiss chemist Ami Argand in the 1780s. In the years that followed, rapid advances culminated in two major transformations in the basic technology of lighting: the development of gas lighting and the invention of arc and incandescent electric lights. The full impact of these developments was not felt until the end of the nineteenth century for gas and the beginning of the twentieth century for electricity. In America, candles and lamps that burned whale oil, lard, lard oil, coal oil, and a host of lesser-known fuels were the most common sources of artificial light throughout most of the nineteenth century.

Argand's central-draft burner increased the oxygen available for combustion of the fuel in oil lamps by conducting an additional current of air to the flame up the center of a tubular wick. The use of a chimney increased the draft and concentrated air at the point of combustion. The light given by the flame of an Argand lamp was much brighter than that of any other lamp of the time. Argand lamps and adaptations of them were available in the United States by the end of the eighteenth century. These lamps were complex devices, consuming more oil than an ordinary oil lamp. Expensive to purchase and to use, they were out of reach for all but the well-to-do.

The existence of a lamp that gave a brighter light contributed to the demand for better light for everyone. The principle of combustion embodied in the Argand central-draft burner, although not totally understood by those who copied the burner and invented new ones, converged with this demand for more light to produce hundreds of inventions for lighting devices. Efforts were made throughout the century to find cheaper, brighter light "for every family," as one lampmaker expressed it. It is this search for the cheapest and best artificial light for the masses that characterizes lighting in nineteenth-century America. Hundreds of patents were issued for lamps, lanterns, burners, wick-raising mechanisms, extinguishers, and other related parts and accessories. And many lighting appliances simply appeared without any formal record of their development.

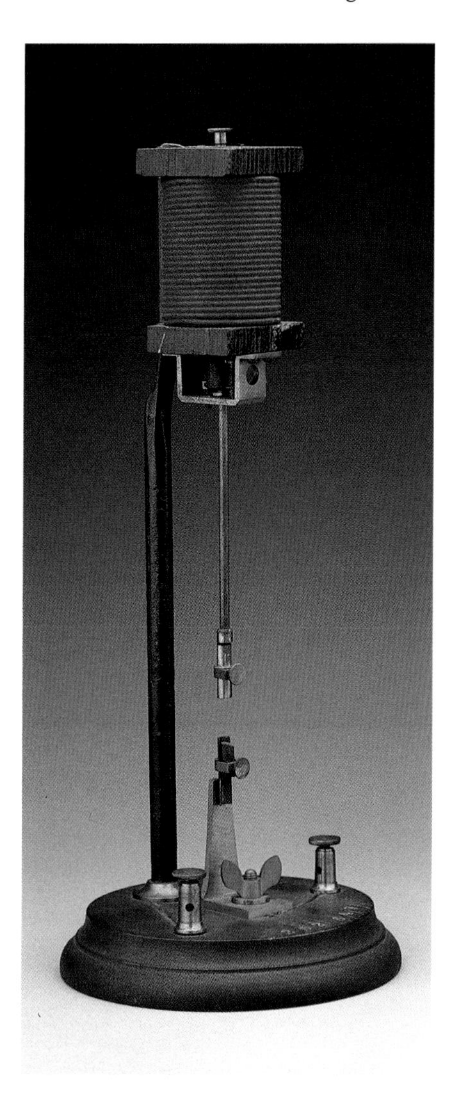

92.
Arc Lamp
Charles F. Brush
Cleveland, Ohio
May 7, 1878
Patent No. 203411

Whale oil was the most common lamp fuel in America at the beginning of the nineteenth century. Simple closed-font single- or multiple-wick tube lamps are so closely identified with this fuel that they are known as whale-oil lamps. These lamps were manufactured by the thousands in glass, brass, pewter, and tinned sheet iron. The growing production of common whale-oil lamps to meet the desire for more light, along with the introduction of the Argand lamp in all its varieties, including astral, sinumbra, and mantel lamps, increased the demand for whale oil. Fluctuations in the whale catch led to lower supplies and higher prices. As whale oil became more expensive, the search for alternate fuels intensified. Any substance that would give an appreciable amount of light when it burned was poured or scraped into lamps of ever-more ingenious design.

One of the first of these new fuels was a mixture of redistilled turpentine and high-proof alcohol patented in 1830 by Isaiah Jennings. This fuel was soon joined by various other combinations of such ingredients. "Burning fluids," as they were called, gave a clear white light and were inexpensive. Unfortunately, they were also very volatile. Even with the change in burner design from the conventional whale-oil burner to the tall single or diverging multiple-wick tubes intended to keep the flame as far as possible from the fuel supply, explosions and fires resulted far too frequently from the use of burning fluids. Many "safety lamps" devised for these fluids continued in use for many years.

Lard, the fat of swine, was a plentiful commodity in nineteenth-century America, especially in farming communities where it was readily available to anyone who raised pigs. This made it an attractive candidate for a lamp fuel. It was problematic, however, in that it was usually solid at room temperature and would not climb a wick under such conditions. Lard could be used in primitive grease lamps, as animal fat had been for centuries, but such lamps did not satisfy the demand for brighter light. Even lard oil, produced by subjecting lard to pressure, was too viscous for easy use in ordinary oil lamps. Between the 1840s and 1860s, several patents were issued to inventors determined to overcome this problem and make cheap, abundant lard an acceptable illuminant. Most of these patents were for small, inexpensive lamps which must have been produced in some quantity as there are large numbers of surviving examples. The majority of these lamps were made of tinned sheet-iron, although brass, glass, and pewter examples are also known. They were meant for use in the homes of farmers and laborers and in the kitchens of the more well-to-do.

The patentees devised various methods to keep the lard in a liquid state and to send it up the wick to the point of combustion. These methods involved the use of heat, usually conducted back from the flame to the lard in the font, to prevent its solidifying. Devices such as canting or tilting fonts kept the distance the liquefied lard had to travel at a minimum, or provided pressure to force the lard up the wick. In essence, the inventors were suggesting ways to burn grease in an oil lamp. These lamps seem to have been peculiarly American devices.

The lamp that was the most successful in employing lard—and actually extended its use to those who could afford more expensive lighting devices—was the solar lamp. This was an Argand lamp rather than a common oil lamp, and it was not originally intended to burn lard. But in 1843 Robert Cornelius of Philadelphia patented an adaptation of the solar lamp designed to burn "lard and other concrete fatty matters." With an all-metal font and a burner that conducted sufficient heat to the fuel, it could burn lard just as easily as whale oil and the other lamp oils of the 1840s. Solar lamps were used in Europe but not ordinarily with lard. The solar lamp of the Cornelius type became very popular in the United States.

The fuel that was to make better lighting possible for the largest number of people in the nineteenth century was coal oil, or kerosene, as it became known in America. The discovery of an abundant supply of the petroleum needed to produce this fuel in the oil fields of Pennsylvania in 1859, and the development of inexpensive lamps to burn it, made kerosene the most popular illuminant in the second half of the century. Although gas lighting had been developed during the eighteenth century, its dependence on a central supply and distribution system limited its usefulness, especially for domestic lighting, to urban areas. In the early years of the nineteenth century, gas lighting was confined primarily to factories, public buildings, and streets. Kerosene lamps could be used anywhere, for they were not tied to a central supply system. Able to burn well in a wide range of lamps, and less expensive as supplies grew, kerosene was affordable for most people.

The first requirement of the new fuel was a burner suitable for the combustion of a carbon-rich substance needing more air to burn. Incomplete combustion produced a smoky flame of diminished brightness. Coal oil could be burned in an Argand lamp, for its central draft and chimney provided additional oxygen to the flame. It did not burn well in the common whale oil lamp or other simple lamps of the period.

The early patents for coal-oil lamps were for burners that could provide more oxygen to the flame. The burner that became the most successful and popular in the early period was the Vienna flat-wick burner. Introduced to the American public in 1857, it appears to have been developed in Vienna and New York simultaneously. Because of a patent dispute, Michael A. Dietz did not receive a patent for his version until 1859. Testimony recorded in the patent hearing revealed that he had been producing such burners since the spring of 1857, a few months before the importation of a similar burner from Vienna. Soon there were dozens of patents issued for burners incorporating adaptations and improvements, and flat-wick burners became standard in inexpensive coal oil or kerosene lamps. These burners had special draft holes and deflectors to increase the air supply to the flame. The usual flat-wick kerosene lamp required a chimney.

Central-draft Argand burners were also used on kerosene lamps. These lamps were usually more expensive than those employing the flat-wick burner. The student lamp, a particularly popular kerosene lamp of the 1870s and later, incorporated both the central-draft burner and the fountain-feed, independent reservoir configuration of early Argand lamps adapted for use with the new fuel. Such student, or study lamps as they were first known, had been in use in Europe with colza oil (a vegetable oil) and whale oil for a number of years. Their appearance in this country can be traced to the lamp patented by Carl A. Kleeman of Erfurt, Prussia, in 1863. Student lamps were at first advertised as "German Study lamps," but with adaptations, they quickly became known by other names, including "American study lamps" and finally, simply "study" or "student lamps." The Kleeman patent was eventually assigned to C.F.A. Hinrichs of New York City, who produced Kleeman's lamp, as well as his own improved "Universal" study lamp.

Kerosene lamps were manufactured in a wide range of forms and styles, from inexpensive hand and table lamps to elaborate floor and banquet lamps, hanging lamps, chandeliers, and wall brackets for use in homes, offices, and public buildings. They were truly the lamps for every family.

By the end of the century, gas lighting—which eliminated the need for wicks, their attendant care, and the filling and cleaning necessary with any oil lamp—had spread to many of the small towns and villages of the country and was a serious challenge to kerosene lighting. The invention of practical electric lighting systems in the 1870s and 1880s ultimately rendered all other forms of illumination obsolete.

ELECTRIC LIGHTS

Based in Cleveland, Ohio, Charles Brush (1849-1929) was one of several American inventors in the 1870s who patented arc lighting systems. Arc lights are very bright and are especially useful for lighting city streets, where they were widely used. The arc in this lamp is formed between two carbon rods held a short distance apart from each other. The current that forms the arc also flows through the solenoid coil at the bottom of the lamp. As the carbon rods burn away, the arc gets longer and less current flows; this reduces the magnetic pull of the solenoid and the rods are allowed to come closer together. The lamp is therefore self-regulating.

Many inventors pursued the quest for a lamp that would produce a softer light

Thomas Edison, after an 1878 demonstration of his phonograph at the National Academy of Sciences, Washington, D.C.

THE DAILY GRAPHIC

Edison, the "Wizard," searchs for platinum. Cartoon, July 9, 1879, *New York Daily Graphic*

Edison's light bulb with patented features

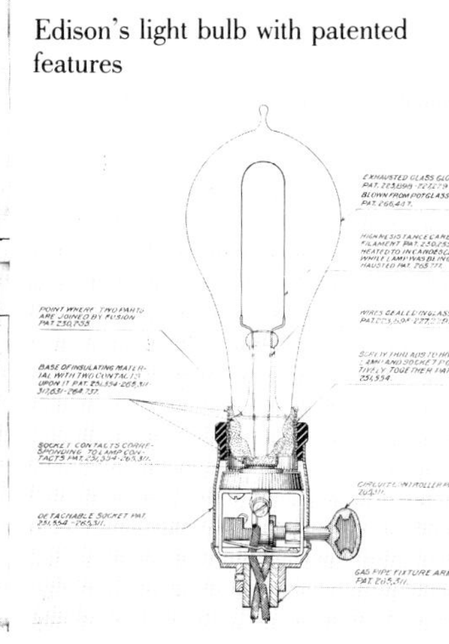

93.
Edison Light Bulbs

1. Early 1880
   Catalog No. U06608U35
2. Early 1881
   Catalog No. 180934
3. Late 1881
   Catalog No. 180935
4. 1886
   Catalog No. 318669
5. 1894
   Catalog No. 318645
6. Japanese bamboo sticks of the kind used for filaments

-1

-2

-3

-4

-5

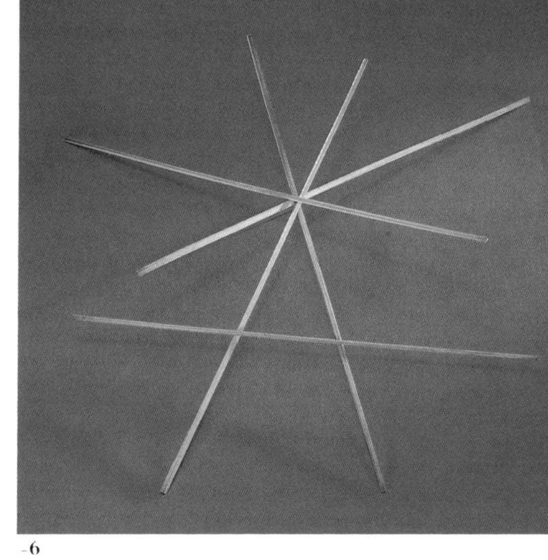

-6

suitable for home use. Edison achieved this goal in 1879 after about a year of concentrated effort. As can be seen in the drawing, there were several critical elements in the light bulb alone. He also developed generators, fuses, meters, conduits, and lighting fixtures for his system. His first large-scale installation was in New York City in 1882.

Thomas Alva Edison was born in the American Midwest in 1847. His formal education was minimal, but he was highly inquisitive and devoted much of his spare time as a child to chemical experiments. As a telegrapher from 1863 to 1869, he learned how to get along with other young men as they shifted from one city to another in search of jobs and excitement. Both his curiosity and his cooperative nature would be important to his career.

In 1876 Edison established his own laboratory at Menlo Park, New Jersey, and in 1886, a much larger operation at nearby West Orange. They can be considered precursors of the industrial laboratories that would become so important in the twentieth century. With the help of these facilities, Edison became America's most prolific inventor. At his death in 1931, he had 1,096 patents to his credit.

Edison's first real money came from his printing telegraph inventions; his contemporary fame (which included the nickname "wizard"), from his invention of the phonograph. But he is now probably remembered most for his contributions to electric lighting. More than a third of his patents (389) were granted in this field.

In the mid-1870s several inventors developed systems for a high-intensity light produced by an electric arc. These were excellent for street lighting, but were too intense for home use. Edison and others began looking for ways to produce a softer light. The most promising method seemed to be by heating some material until it glowed, becoming "incandescent."

After more than a year of experimentation, Edison found a solution late in 1879: a carbon filament in an evacuated glass envelope. His first successful filaments were produced by baking thread in the absence of air until nothing was left but carbon. In December 1879 he demonstrated some lamps that used carbonized cardboard. But early in 1880 he found what he considered the "ultimate" natural material: bamboo, which he imported from Japan. But further work needed to be done. As can be seen in the drawing, a number of innovations were necessary for the light bulb alone. How he varied some of these in succeeding years can be seen in the examples on page 59.

1. Early 1880. One of Edison's first bamboo filament lamps, with his first form of screw base. Note how the bamboo is wider at each end so that it can be grasped by the platinum clamps and so good conduction will occur.

2. Early 1881. Copper plating has replaced the expensive and awkward clamps. The base has been changed so that screwing the light bulb in place produces pressure on the contacts.

3. Late 1881. Plaster has replace wood in the base, which now has taken a form that remains to the present day. The plaster was given a ridge which could be gripped, so that one would not hold the glass while screwing it in and thus risk breaking the seal.

94

95

4. 1886. Carbon paste has replaced the copper plating. This also meant that the ends of the bamboo no longer had to be flared. Note also that with a stronger bond between the glass and plaster, the ridge had been left off the base in 1884.

5. 1894. Techniques for producing filaments using carbonized squirted cellulose were developed in Europe in 1888. But the Edison company, now a part of General Electric, continued to use bamboo until 1894. This is one of their first squirted cellulose lamps. Because its specific conductivity is greater, the filament had to be made longer.

Edison's early light bulbs produced about 1.7 lumens per watt, though over time some of the carbon would be deposited on the glass and lower the efficiency. Carbon lamps, improved to an efficiency of about 3.4 lumens per watt, continued to be manufactured until 1918. Today's incandescent lamps, with tungsten and other metalized filaments, operate at efficiencies ten times and more over the 1881 model.

97

ORIGINAL AND ONLY GENUINE

Saint Germain,

—OR—

German Study or Office Lamp.

C. A. KLEEMANN'S PATENT.

Prize Medal of the American Institute Exhibition, Oct. 26, 1867. Diploma of the "Pennsylvania State Fair," 1866.

First Diploma of the Kings County Fair, 1872.

LIGHT! MORE LIGHT!!

Best, Safest, Handsomest, Most Economical. No Odor, no Smoke. Pure, Brilliant, Unwavering Light, very agreeable to the Eye. None Genuine without my name on Chimney-Holder. This Lamp is all of METAL, Easily Managed.

Trade Mark on Chimney Holders.

"Patented Mar. 10, 1863, Reissued Dec. 30, 1873, C. A. Kleemann. C. F. A. Hinrichs, New York,

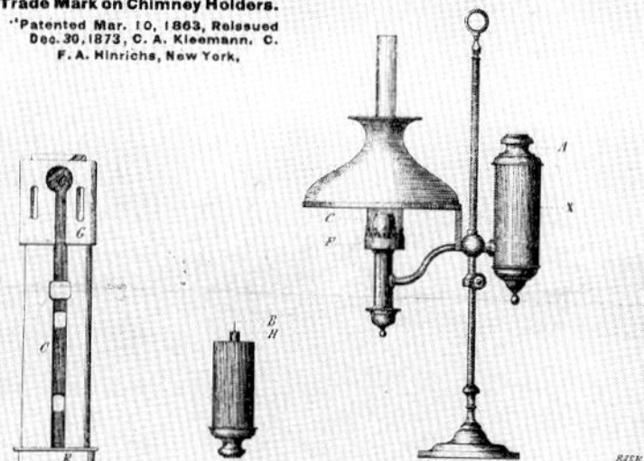

DIRECTIONS FOR USE.

To fill the Lamp, take out the Holder, *A*, invert it and pour in the Oil till it reaches the valve; then pull up the valve by means of the wire *B*, invert it, holding it above the holder *X* so that any oil which may escape drops into this Holder; replace it in the Holder *X*. To put on the wick, take off the Chimney holder *F*, take out the Cylinder *C*, take out the smallest Cylinder *D*, tie the wick at the base of Cylinder *D*, then replace the Cylinder *D* in the large one *C*, taking care to push it down as far as the point *E*. Replace everything as before, the large Cylinder *C* having the Brass catches *G* up and the Ring *E* down. To raise the wick turn the Chimney-holder.

This Lamp gives a very superior and steady light, and with ordinary care will emit neither smell nor smoke. One-twelfth, or one eighth of a heavier oil, Sperm, Lard or Olive, mixed with Kerosene, makes the best and safest oil.

Testimonials have been given by highest authority as to its safety against explosions.

Besides the above pattern Lamps, in Brass, as well as in German Silver, I also have Lamps with two burners or Double Lamps; also, Bracket or Side Lamps, and Double Hanging Lamps or Keroseniers, all with the same St. Germain patent burners.

The wick should be trimmed regularly. If a crust has formed, do not disturb it, but only remove any little point or unevenness that may occur; do not use the scissors unless the wick, through uneven draft, should have coaled or charred unevenly. By this method you will have an even flame, and the wick will last much longer than when cut frequently. If your Lamp should make a humming noise, which is caused by the shank of the chimney being of the wrong length, raise the chimney slightly, or change it for one with a longer shank.

Use kerosene or spirits in place of water for cleaning chimneys. The brass part of the Lamp may be cleaned with Vienna lime and kerosene, and polished with rouge.

N. B.—Parties returning lamps for repairs will please see that every lamp is properly labelled, with name and address, and also with the complaint.

As there are now being offered to the public, worthless articles, purporting to be chimneys and wicks designed to be used on my ST. GERMAIN LAMP, but really intended to injure its sale, particular attention is called to use a proper sized chimney and wick.

Size of chimney for No. 1 Lamp, full length, about 10 inches, shank or wide part, 2 inches high, opening in the neck not less than [illegible] inch clear, diameter 1[illegible]. The chimney for No. 2 Lamp has to be about the same size, except the diameter of shank should only be 1[illegible] inch to fit in the holder. Wicks have to be large enough to fit close to the wick holder. A too large wick will be apt to crease or fold and give an uneven circle of light, while burning, besides such a wick will draw the oil quicker than it can be consumed by the flame.

Advertisement for Kleemann's patented student lamp, about 1873

96

PURE LIGHT. 257

Where may be found a good assortment of

Hang, Side and Bracket Generator Lamps,
Hang and Side Camphene Lamps,
Hang & Side Solar do.
Girandoles,
Chandeliers,
Astral Generator Lamps,
Plated Glass Lamps,
Night Lamps,
Peg do.
Study Shades,
Patent Conductor Lanterns,

FLUID & CAMPHENE.
CHANDELIERS, LAMPS, &C.
J. F. DODGE.

Together with a good assortment of

Fluid Lamp-tops, Brass and Britannia,
Hall and Hand Lanterns,
Bracket Lamps,
Study do.
Glass Chimnies,
Fluid Wicks,
Camphene Wicks,
Solar Wicks,
Tin Cans,
Brass Chain,
Glass Drops,
Patent New England Spring Bottom Lanterns,

J. F. DODGE,

AT 36 UNION STREET, BOSTON, MASS.,

Manufactures and offers at Wholesale and Retail,

PORTABLE BURNING FLUID,

AND

CHANDELIERS, LAMPS, GIRANDOLES, LANTERNS, &C. &C.

FOR USING THE SAME,

Adapted to the lighting of Churches, Halls, Factories, Stores, Railroad Stations, Cars, Dwellings, Workshops, Carriages, Streets, Ships, &c.

ALSO, CAMPHENE OF EXTRA QUALITY,

And a constant supply of

BLAKE'S PATENT

Arm or Conductor Lantern,

An article which should be in the possession of every person who wishes to carry his own light, and at the same time have both hands at liberty.

N. B.—Any article in the above line made to order at short notice.

Blake's patented lantern advertised for sale by Dodge's Lamp Store, about 1852

94.
Lard Lamp
Delamar Kinnear
Circleville, Ohio
February 4, 1851
Patent No. 7921

95.
Foot Warmer
Francis Arnold
Haddam, Connecticut
April 11, 1854
Patent No. 10769

96.
Railroad Lantern
Philos Blake
New Haven, Connecticut
January 13, 1852
Patent No. 8650

97.
Student Lamp
Carl A. Kleeman
Erfurt, Prussia
March 10, 1863
Patent No. 37867

Steam

No other mechanical device played such an essential role in determining the direction of nineteenth-century American civilization as did the steam engine. This prime mover freed mankind from dependence on human or animal muscle power and from the vagaries of natural forces such as wind and water. The steam engine could operate unfettered by any conditions imposed by either weather or geography. Because it could be used to power just about any type of tool or machine necessary to the manufacture of goods, it became an essential part of the economic growth of the nation. Steam engines could be erected just about anywhere and so made it possible to bring industry and its attendant progress into every community. Engines came in all sizes and were basically easy to operate. They required little more than water and fuel to make them perform. It was their rather voracious need for fuel that influenced the way in which they developed.

The steam engine itself burned no fuel. Steam, which resulted from water being heated in a closed vessel—a boiler—provided the power to drive the engine which in turn did the actual work. Because the once-abundant supply of wood had been so depleted by mid-century, coal had all

98.
Valve Mechanism
Mathias L. Jacquemin
Council Bluffs, Iowa
September 23, 1879
Patent No. 219950

99.
Cut-Off Valve Gear for Steam Engine
William G. Pike
Philadelphia, Pennsylvania
November 20, 1866
Patent No. 59777

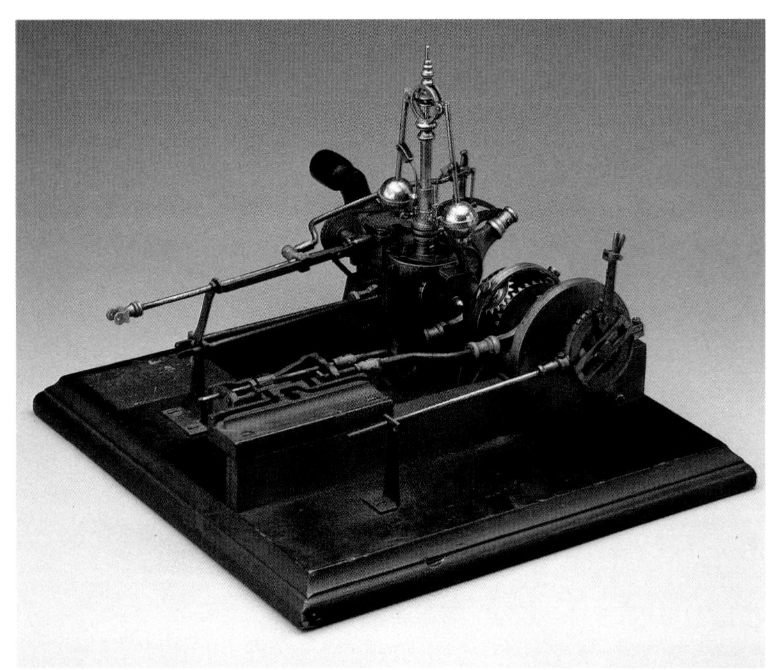

98

99

Steam engine in Council Bluffs, Iowa, 1850

100

100.
Dog-Powered Treadmill
Frederick H. Traxler
Dansville, New York
April 23, 1878
Patent No. 202679

101.
Windmill
Jesse Benson
Champaign County, Ohio
November 12, 1878
Patent No. 209853

102.
Windmill
Henry H. Bevil
Indianapolis, Indiana
April 6, 1880
Patent No. 226265

Windmill drawing water for cattle in Wyoming

101

102

but replaced it as the primary fuel. Found in great abundance in many areas, coal was an ideal alternative. The cost of mining and transporting it, however, made steam engines more expensive to operate. This economic reality—the cost of fuel—was critical in the further development of the steam engine.

Of the various components improved during the nineteenth century, perhaps none was more significant than the valve gear. This critical element in all engines regulated when, and how much, steam entered the cylinder. If the valve admitted more than was needed, or not enough, causing the engine to lag, both steam and coal were needlessly wasted, for the engine was not operating efficiently. Inventors who aimed to improve valve gear all sought to meter the amount of steam being used more carefully, mainly by making valve action more responsive to the workload. It was not uncommon for an engine builder to encourage the sales of his engine by charging no more for it than what the user would save in the price of coal!

WIND

While the steam engine fostered industrial development in American cities and towns, the wind engine, or windmill, permitted increased agricultural development on the most arid land. By using the free natural force of the ever-present wind, the windmill pumped life-sustaining water from beneath the earth's surface. Although windmills were found throughout the country, their greatest use was on the plains of the Midwest. The idea of the windmill was not original to the United States, but by the mid-nineteenth century there developed what would become known as the "American-style" windmill. These usually consisted of a multi-bladed lightweight rotor, often a single regulating vane to the rear, and a geared mechanism for converting wind energy to a usable form, all mounted on a framework tower.

Patents centered around imaginative methods for extracting the greatest amount of energy, and therefore work, from the slightest breeze. This was accomplished through cleverly designed blades of wood as well as sheet metal. Of equal importance was the mechanism that automatically governed or regulated the mill's operation. It was this feature that set American windmills apart from all others. Countless devices were patented for keep-

103

104

105

106

103.
Hot Air Engine
A.S. Lyman
New York, New York
February 28, 1854
Patent No. 10576

104.
Gas Engine
Nicholaus A. Otto
Deutz, Germany
May 30, 1876
Patent No. 178023

105.
Electric Generator
Elihu Thomson
Philadelphia, Pennsylvania
October 5, 1880
Patent No. 233047

106.
Electric Generator
Hiram S. Maxim
Brooklyn, New York
November 2, 1880
Patent No. 233942

ing the blades faced into the wind and for moving them aside when the wind became too strong. Other patents pertained to improvements to the machinery that converted the rotary motion of the blades into the reciprocating motion necessary for pumping.

Although the wind engine was used mainly for pumping, it also was employed to operate small cereal-grinding mills. A farm or ranch equipped with a windmill could extract energy—and therefore the power to do work both day and night—from the unseen natural force that moved almost constantly.

Hot Air

Hot-air engines devised in the early nineteenth century seemed to offer all the advantages of the steam engine while avoiding the danger and inconvenience. Using an operating cycle that relied on the expansive power of air heated in a closed cylinder and the contractive force of cold air, these engines were not totally unlike the steam engines they were intended to supplant. In appearance and application they were quite alike. But rather than heating a boiler to produce steam, the cylinder of the engine itself was heated. The engine would continue to run as long as the cylinder continued to be heated. Extremely inefficient and low powered, hot-air engines were inappropriate for use in large-scale applications that might be needed in a factory. In spite of their inherent drawbacks, however, they were a commercial success. Easily providing the limited power needed by the average American farm, they were ideal for pumping water or running a small machine. Fired with coal or the wood scraps found on most farms, they placed little burden on the farmer who used one.

Internal Combustion

By the late 1880s the internal combustion engine had developed to the point where it was in direct competition with the steam engine. Like the steam engine, its capabilities for providing power were unlimited, and it had other advantages. Utilizing the expansion of burning gases in a closed cylinder, power was extracted from fuel burned right in the engine. There was no need for a boiler; fuel was either a flammable gas or vaporized liquid. Though the fuel could be dangerous in itself, it did not pose the lethal threat of boiler explosions, as steam engines did.

Internal combustion engines offered power comparable to steam, along with economy and smooth operation. Early examples with horizontal cylinders and large flywheels were similar in appearance to steam engines. Patents covered all aspects of the engines: improvements to ignition systems, carburetion, pistons, cylinders, and valves. Although they could be started instantly, were self-contained, and did not consume fuel when they were at rest, their greatest potential went untapped until the twentieth century, when they would be adapted for use in the automobile.

Electricity

The principle of converting motive power into electricity became known in the 1830s. Several small, inefficient generators were designed, using permanent magnets to produce the magnetic field. In 1866 the self-excited dynamo, which used electromagnets instead of permanent magnets for the field, was invented independently in England and Germany. A late example was shown at the Philadelphia Centennial Exposition in 1876, triggering much activity in the United States.

Known for his automatic machine gun and other inventions, Hiram Maxim (1840-1916) was one of many who designed improved electric generators. Notice that the magnetic field (in the several iron bars) is produced by electrified coils, while the armature rotates inside.

One of the best-known American electrical inventors of the nineteenth century was Elihu Thomson (1853-1937), cofounder of the Thomson-Houston Company that merged with Edison's company in 1892 to form General Electric. His very distinctive self-excited dynamo was widely used in electroplating and lighting systems.

PATENTED PASTIMES

★ ICE SKATING

★ STRINGED INSTRUMENTS

"Central-Park Winter. The Skating Pond," lithograph by Currier and Ives, 1862, Library of Congress

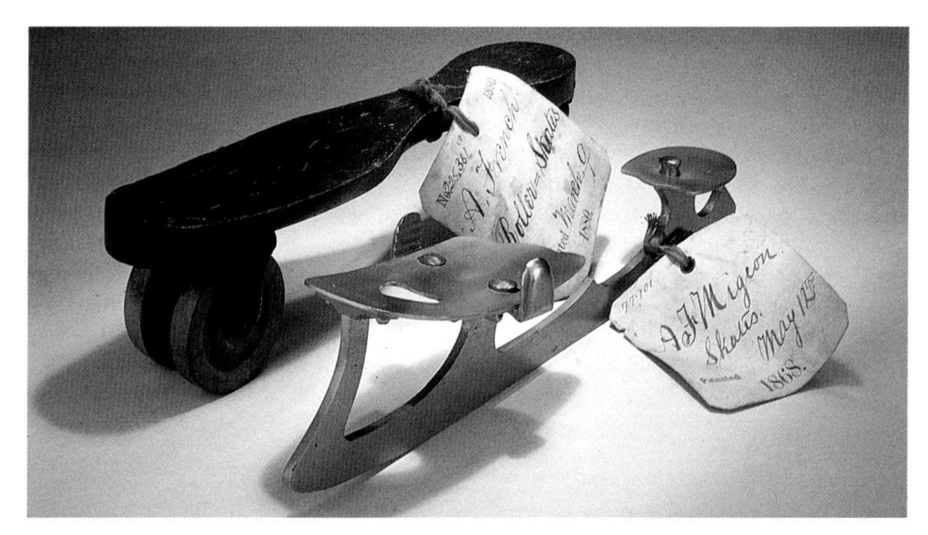

107.
Roller Skates
Andrew French
Philadelphia, Pennsylvania
March 9, 1880
Patent No. 225361

108.
Ice Skates
Achille F. Migeon
Wolcottville, Connecticut
May 12, 1868
Patent No. 77901

ICE SKATING

Recreation, like nearly every aspect of American life, was transformed in the nineteenth century. The popularity of sports and leisure activities of all kinds grew tremendously after 1860. Urbanization and industrialization threw together ever greater numbers of people with more time and more money, and with physical and psychic energy in need of release. Ice skating, among the first popular athletic activities, was quickly commercialized and promoted. It met new needs and desires for healthful recreation, inexpensive entertainment, and social interchange, and it fit the democratic ideal as a sport open to all. Ice skating was one of the first strenuous physical pastimes women enjoyed in public—part of its appeal to men as well.

"Glacariums," or indoor ice rinks, made possible by the invention of artificial icemaking equipment and gas lighting, increased the hours available for skating and lengthened the season. This victory over nature enhanced the level of comfort and gentility, and removed a real fear of falling through the ice.

A dramatic rise in sales of skating equipment, instruction manuals and skating club memberships between 1860 and 1890 testifies to the popularity of this social, yet individualistic, sport. Skates, mainly European imports up to 1860, were sold in hardware stores, at the indoor and outdoor rinks, and increasingly at specialty stores. Wooden skates with metal blades and leather straps were common throughout most of the century. Hundreds of patents were filed on variations and improvements for women's skates, beginners' skates, faster skates, safer skates. Ladies' models were made with leather cups to secure the heels. To help beginners, I.F. Blondin patented an innovative design with leg supports for weak ankles. All-metal skates secured with clamps, called club skates, were manufactured from the 1870s on. This type displaced the wooden stock skate, which often split, and dominated sales until the advent of the boot skate, which has remained fashionable since the turn of the century. All these innovations reflected and contributed to the popularity of the sport.

The skating craze faded in the 1890s as fads customarily will. Competition from new sports like basketball and inventions such as the safety bicycle put skating in the background. Yet we can still see that ice skating's union of social needs with technical innovations and commercialization exemplifies the creative impulses of the nineteenth-century. These improvements in sporting equipment, educational devices, and games do not highlight inventions that changed the world, but an aggregate, characteristic American pursuit of novelty.

One of several similar hunting and fishing scenes produced by various lithographers in the 1860s and 1870s

109

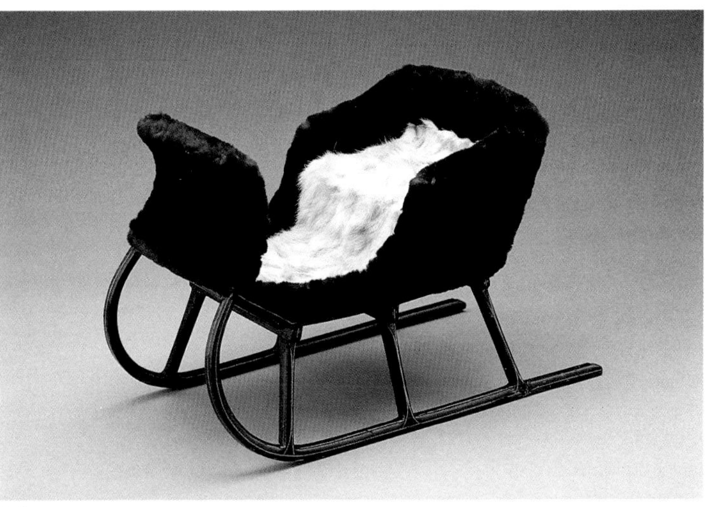

110

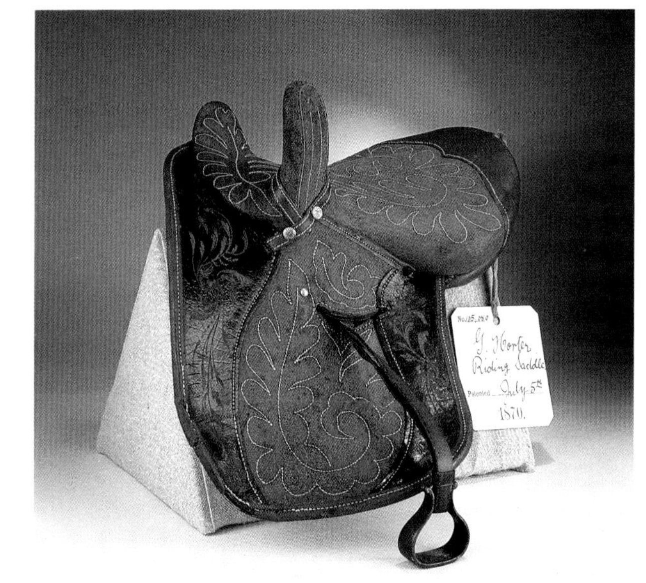

111

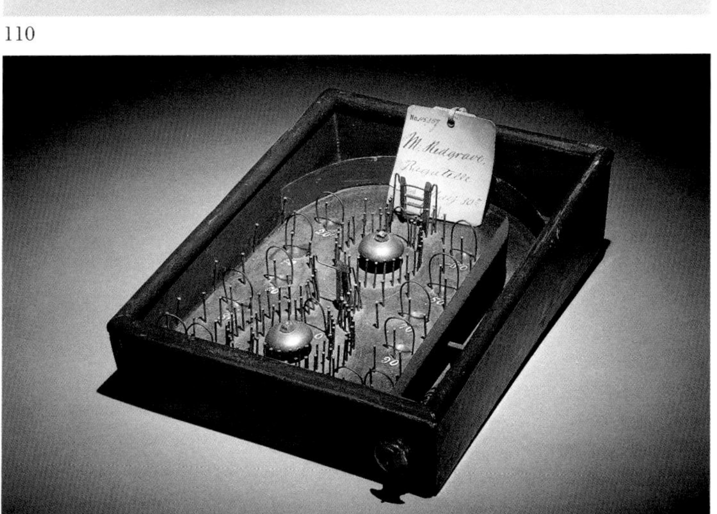

112

109.
Fishing Reel
Charles W. MacCord
Weehawken, New Jersey
February 10, 1874
Patent No. 147414

110.
Sleigh with Fur
Martin S. Davis
Toledo, Ohio
October 19, 1880
Patent No. 233331

111.
Saddle
George Horter
New Orleans, Louisiana
July 5, 1870
Patent No. 105080

112.
Bagatelle
Montague Redgrave
Cincinnati, Ohio
May 30, 1871
Patent No. 115357

Stringed Instruments

William Sidney Mount (1807-1868) is best known as an important American genre painter. Mount's portraits, landscapes, and scenes document daily American life, primarily that of his native Long Island, New York. Possessed of a keen curiosity, Mount invented a steamboat paddlewheel, a two-hulled sailboat, a painting studio on wheels, and his hollow-back violin, named the "Cradle of Harmony." A talented country fiddler, Mount assembled a large volume of dance tunes—waltzes, polkas, jigs, and schottisches—adapted for fiddling. The desire for a more powerful sound in a dance hall probably led to his inventing the new instrument. In his 1852 patent Mount claims to "have invented a new and improved mode of constructing violins and other stringed musical instruments by which a greater strength of the parts is secured with a greater lightness of the material composing the instrument, and at the same time a superior quality and greater quantity of tone and sound are obtained." He believed that a concave shape and a short soundpost would result in a fuller, richer, more powerful tone.

Mount first wrote to the United States Patent Office on January 2, 1852, and was informed that he had to include a model of his patent. He delivered this instrument on February 23, and was granted Patent No. 8,981, dated June 1, 1852. The patent document protected his exclusive manufacturing and vending rights for fourteen years. Mount displayed his instruments at the New York Crystal Palace in 1853, demonstrating the hollow-back model in person. He sought endorsements of his violin from contemporary musicians, but never manufactured his invention in great quantity.

Sewall Short of New London, Connecticut, had another approach to the attainment of greater sound in instruments of the violin family. In his Patent No. 10,867 (May 2, 1854), he states that by the application of his invention to a violin, violoncello, bass viol, or instrument of similar character "the vibrations of the latter are greatly increased and the tone and power of the instruments much improved." His solution was simply to fit a metallic horn or trumpet to a hollow wooden neck of a normal violin. This model, which he submitted with his patent application, has a brass horn attached to a violin bearing the label of the French violin maker, Derazey. With this instrument, a musician could presumably project his artistry in a focused direction.

While the "amplified horn" violins of the Englishman John Stroh became popular in recording studios at the turn of the twentieth century, few violins of Sewall Short's design have survived, and his idea apparently did not win approval from concert artists or violin manufacturers.

113

114

113.
Violin
William S. Mount
Stoney Brook, New York
June 1, 1852
Patent No. 8981

114.
Violin with Horn
Sewall Short
New London, Connecticut
May 2, 1854
Patent No. 10867

Rows of cases with models in the Patent Office Model Hall, from a stereoview card, about 1870

Objects in Traveling Exhibition to Japan
July 1–December 3, 1989

■ No	■ Invention	■ Inventor	■ Date	■ Patent No.
1	Locomotive	Ross Winans	July 29, 1837	No. 305
2	Locomotive	Matthias W. Baldwin	August 25, 1842	No. 2759
3	Locomotive	Andrew Cathcart	October 23, 1849	No. 6818
4	Automobile	George B. Selden	November 5, 1895	No. 549160
5	Ford Model T Runabout		1926	No. 333777
6	Columbia Bicycle		1886	No. 307217
7	Tricycle	Otto Unzicker	June 4, 1878	No. 204636
8	Tricyle	Charles Hammelmann	March 2, 1880	No. 225010
9	Tricycle	Francis Fowler	February 3, 1880	No. 224165
10	Vessel (Replica)	Abraham Lincoln	May 22, 1849	No. 6469
11	Steamboat Steering Gear	Frederick E. Sickels	May 10, 1853	No. 9713
12	Steam Steering Apparatus	Frederick E. Sickels	July 17, 1860	No. 29200
13	Paddle Wheel	Fletcher Felter	November 8, 1854	No. 11992
14	Edison Phonograph		About 1878	No. 320551
15	Telegraph	Samuel F.B. Morse	May 1, 1849	No. 6420
16	Morse Telegraph (Replica)		1844	No. 200002
17	Printing Telegraph	Thomas Edison	July 1, 1873	No. 140488
18	Fire Alarm	Moses Farmer & William Channing	May 19, 1857	No. 17355
19	Alarm Telegraph	Edwin Rogers	June 14, 1870	No. 104357
20	Bell Telephone		About 1898	No. 181852
21	Telephone	Alexander G. Bell	March 7, 1876	No. 174465
22	Telephone	Thomas Edison	September 23, 1878	No. 208299
23	Edison Telephone		About 1877	No. 314896
24	Long Distance Telephone	Anthony C. White	November 1, 1892	No. 485311
25	Machine for Making Paper Bags	Margaret E. Knight	October 28, 1879	No. 220925
26	Printing Press	Richard M. Hoe	July 24, 1847	No. 5199
27	Printing Press	William H. Golding	December 2, 1873	No. 145101
28	Printing Press	Thomas C. Kenworth & Archibald McGregor	May 7, 1878	No. 203465
29	Hand Printing Press	James N. Phelps	November 2, 1858	No. 21980

■ No	■ Invention	■ Inventor	■ Date	■ Patent No.
30	Type Casting Machine	David Bruce, Jr.	November 6, 1843	No. 3324
31	Camera	William Southworth	June 17, 1862	No. 35635
32	Camera	John Stock	July 5, 1859	No. 24671
33	Remington No. 1 Typewriter		About 1878	No. 181132
34	Typewriter	Hans R.M.J. Hansen	April 23, 1872	No. 125952
35	Typewriter	C. Latham Sholes & Matthias Schwalbach	September 19, 1876	No. 182511
36	Hansen Typewriter		About 1872	No. 181005
37	Animal Trap	George F. Lampkin	July 9, 1872	No. 128802
38	Animal Trap	A. A. Fradenburg	May 22, 1866	No. 54885
39	Cotton Seed Planter	A. W. Washburn	March 26, 1866	No. 14529
40	Traction Engine	John E. Praul	November 4, 1879	No. 221354
41	Reaper	William N. Whiteley	November 13, 1877	No. 197192
42	Grain Binder	Harvey R. Ingledue	February 3, 1885	No. 311492
43	Plow	Augustus G. Christman	January 20, 1880	No. 223666
44	Planter	Charles G. Everet	July 13, 1880	No. 229985
45	Corn Planter	George W. Brown	May 29, 1883	No. 278497
46	Cultivator	Philip F. Wells	January 24, 1882	No. 252637
47	Mowing Machine	James Herva Jones	June 23, 1891	No. 454741
48	Cultivator	George B. Davison	January 26, 1892	No. 467660
49	Vegetable Sorter	John H. Heinz	February 4, 1879	No. 212000
50	Flax Seed Separator	Jeremiah A. Wade	September 30, 1879	No. 220211
51	Loom	George Crompton	September 7, 1869	No. 94571
52	Knitting Machine	Richard Walker	December 5, 1839	No. 1421
53	Eli Whitney Cotton Gin	Courtroom model	About 1800	No. T.8756
54	Carpet Loom	Erastus B. Bigelow	May 30, 1876	No. 177920
55	Carding Machine	Hiram Houghton	April 21, 1857	No. 17094
56	Design for Carpet	Henry G. Thompson	August 7, 1860	No. 1306
57	Spinning Machine	Charles Danforth	December 12, 1854	No. 12055
58	Sewing Machine	Job A. Davis	February 21, 1860	No. 27208

■ No	■ Invention	■ Inventor	■ Date	■ Patent No.
59	Sewing Machine	James Perry	November 23, 1858	No. 22148
60	Sewing Machine	David W. Clark	July 5, 1858	No. 19015
61	Sewing Machine	George Hensel	July 12, 1859	No. 24737
62	Sewing Machine	Allen B. Wilson	June 15, 1852	No. 9041
63	Sewing Machine	David W. Clark	January 19, 1858	No. 19129
64	Sewing Machine	William O. Grover	May 27, 1856	No. 14956
65	Sewing Machine	David W. Clark	August 31, 1858	No. 21322
66	Sewing Machine	Isaac M. Singer	August 12, 1851	No. 8294
67	Pocket Watch	Charles E. Jacot	September 27, 1864	No. 44493
68	Pocket Watch	Daniel Azro A. Buck	December 16, 1879	No. 222658
69	Gear-Cutting Machine	John A. Peer	July 21, 1874	No. 153370
70	Candy-Making Machine	Thomas and George Mills	February 14, 1871	No. 111765
71	Pill-Coating Machine	William Cairnes	February 16, 1875	No. 159899
72	Adding Machine	Jabez Burns	August 24, 1858	No. 21243
73	Adding Machine	Gustavus Linderoos	June 24, 1873	No. 140146
74	Wood Plane	Henry B. Price	June 17, 1879	No. 216698
75	Machine for Treeing Boots & Shoes	Willard Comey	March 5, 1878	No. 200979
76	Grinding Machine	Joseph L. Hayden	January 13, 1880	No. 223507
77	Portable Forge	Charles Hammelmann	January 10, 1882	No. 252103
78	Combination Tool	John Graham	November 5, 1867	No. 70547
79	Stove	Stephen Culver	December 1, 1868	No. 84537
80	Oven	Alexander White	February 8, 1870	No. 99613
81	Rocking Chair with Fan	Haro J. Coster	May 8, 1860	No. 28159
82	Sofa and Bed Combined	Julius Werner	September 12, 1871	No. 118994
83	School Desk	Joseph Ingels	May 10, 1870	No. 102941
84A	Folding Chair	Claudius & Nicholas Collignon	September 29, 1868	No. 82494
84B	Collignon Folding Chair		About 1870	No. 307707.4
85	School Desk	Charles Perley	May 24, 1859	No. 24151
86	School desk	Sylvanus Cox & William Fanning	January 21, 1873	No. 135089

■ No	■ Invention	■ Inventor	■ Date	■ Patent No.
87	Playpen	Hiram J. Parker	November 2, 1875	No. 169471
88A	Maytag Washing Machine		1915	No. 330502
88B	Washing Machine	A. J. Stafford & S. Crossman	January 9, 1866	No. 51977
89	Washing Machine	William Wheeler	November 28, 1854	No. 12012
90	Refrigerator or Ice Box	Adam Heinz	April 1, 1879	No. 213751
91-1	Clothespin	Dexter Pierce	May 28, 1858	No. 20364
91-2	Clothespin	Jeremiah Greenwood	November 15, 1864	No. 45119
91-3	Clothespin	Henry W. Sargeant, Jr.	April 11, 1865	No. 47223
91-4	Clothespin	T. L. Goble	December 18, 1866	No. 60627
91-5	Clothespin	David M. Smith	April 9, 1867	No. 63759
91-6	Clothespin	A. L. Taylor	April 7, 1868	No. 76547
91-7	Clothespin	Henry Mellish	October 17, 1871	No. 119938
91-8	Clothespin	Vincent Urso & Benjamin Charles	August 12, 1873	No. 141740
91-9	Clothespin	Richard B. Perkins	February 20, 1883	No. 272762
92	Arc Lamp	Charles F. Brush	May 7, 1878	No. 203411
93-1	Edison Light Bulb		Early 1880	No. U06608U35
93-2	Edison Light Bulb		Early 1881	No. 180934
93-3	Edison Light Bulb		Late 1881	No. 180935
93-4	Edison Light Bulb		1886	No. 318669
93-5	Edison Light Bulb		1894	No. 318645
93-6	Japanese bamboo sticks of the kind used for filaments			
94	Lard Lamp	Delamar Kinnear	February 4, 1851	No. 7921
95	Foot Warmer	Francis Arnold	April 11, 1854	No. 10769
96	Railroad Lantern	Philos Blake	January 13, 1852	No. 8650
97	Student Lamp	Carl A. Kleeman	March 10, 1863	No. 37867
98	Valve Mechanism	Mathias L. Jacquemin	September 23, 1879	No. 219950
99	Cut-Off Valve Gear for Steam Engine	William G. Pike	November 20, 1866	No. 59777
100	Dog-Powered Treadmill	Frederick H. Traxler	April 23, 1878	No. 202679
101	Windmill	Jesse Benson	November 12, 1878	No. 209853

■ No	■ Invention	■ Inventor	■ Date	■ Patent No.
102	Windmill	Henry H. Bevil	April 6, 1880	No. 226265
103	Hot Air Engine	A.S. Lyman	February 28, 1854	No. 10576
104	Gas Engine	Nicholaus A. Otto	May 30, 1876	No. 178023
105	Electric Generator	Elihu Thomson	October 5, 1880	No. 233047
106	Electric Generator	Hiram S. Maxim	November 2, 1880	No. 233942
107	Roller Skates	Andrew French	March 9, 1880	No. 225361
108	Ice Skates	Achille F. Migeon	May 12, 1868	No. 77901
109	Fishing Reel	Charles W. MacCord	February 10, 1874	No. 147414
110	Sleigh with Fur	Martin S. Davis	October 19, 1880	No. 233331
111	Saddle	George Horter	July 5, 1870	No. 105080
112	Bagatelle	Montague Redgrave	May 30, 1871	No. 115357
113	Violin	William S. Mount	June 1, 1852	No. 8981
114	Violin with Horn	Sewall Short	May 2, 1854	No. 10867

Hinkley and Williams 4-4-0 Locomotive built in 1870

Icons of Invention: American Patent Models
was designed by
The Watermark Design Office and
printed by Garamond/Pridemark Press, Inc.
The text was set in 10.5 Bodoni Book by
AGT/Unicorn Graphics and
printed on LOE Dull White 80 lb. text and
LusterKote 10 pt. cover.